The Nuclear Force

The Force Which Binds Quarks Into Protons and Neutrons

By

Shelton W. Riggs, Jr.

The Nuclear Force

**The Force Which Binds
Quarks Into
Protons and Neutrons**

Copyright © 2012 by Shelton W. Riggs, Jr

All rights reserved. No part of this book may be used or reproduced by any means, graphic, electronic, or mechanical, including photocopying, recording, taping or by any information storage retrieval system without the written permission of the publisher except in the case of brief quotations embodied in critical articles and reviews.

ISBN-13: 978-1500765385

ISBN-10: 1500765384

This book is dedicated to
all who are curious about
Nuclear Forces

About The Author

Shelton W. Riggs, Jr. earned undergraduate (University of Texas) and graduate (Vanderbilt) degrees in both Physics and Mathematics.

Professionally, he has consulted as both a hardware and software design engineer to numerous Fortune 500 companies for a wide range of scientific applications. He helped solve several scientific problems for US Army, Air Force and Navy.

Other interests include theoretical physics including quantum mechanics, relativistic mechanics and theoretical mathematics (especially the mystery of prime numbers).

Hobbies include dancing, karaoke, juggling, playing keyboards, writing songs, and writing poetry.

Other Works By Author

The Scientific Theory of God – A bridge Between Faith and Physics provides the reader with basic scientific understanding, interpretation, clarification and answers about concepts and beliefs associated with a Supreme Being. These ideas are developed and based on current theory and the standard model of physics. This new basis has revealed surprising relationships between the scientific definitions of both God and man. A model for the behavior of living matter (bioenergy) has been extended to include the behavior of human beings in terms of perception, decision and action. These concepts combined with the operation of short and long-term memory explain both human consciousness and how the mind controls the body. This model also includes how any desired behavior (provided it does not go against survival) may be achieved. This book offers a scientific creation theory and shows how it is compatible with both the big bang as well as evolutionary theory.

Nature of the First Cause – The Discovery of What Triggered the Big Bang contains the formal scientific theory of how the universe got started. It lays down the mathematical foundation for the creation theory put forth in "The Scientific Theory of God" book. It resolves the asymmetry problem of physics. It solves the two main cosmological problems by identifying both dark energy and dark matter. This theory predicts the correct order of magnitude for the number of galaxies and stars in the universe revealed by the Hubble ultra deep field results. It uncovers two entangled parallel worlds consisting of negative antimatter and positive matter. It explains the accelerated expansion of both matter and negative antimatter. It predicts the distance between matter and negative antimatter to be the Schwarzschild diameter of the expanding universe.

An Alternate Lorentz Invariant Relativistic Wave Equation offers an invariant form which differs from both the Dirac equation as well as the Klein-Gordon equation. Unlike, Schrodinger's non-relativistic wave equation, both the Dirac as well as the Klein-Gordon equation predict wave functions which do not collapse when applied to free systems at rest. On the other hand, Schrodinger's equation predicts wave functions that do collapse when applied to free systems at rest. The alternate relativistic wave equation offered by the author follows Schrodinger's philosophy that if a free system is at rest, then it is a particle with a collapsed wave function.

The Origin of the Planck Mass, Planck Length and Planck Time presents solutions of a system composed of two identical photons which are trapped in each other's gravitational field. The solution applies to any pair of identical particles having zero rest mass. Two solutions were derived. One solution was found by treating two photons as point particles. The quantum mechanical solution came about by treating the two photons as waves. In both solutions, the predicted distance between photons was found to be proportional to the Planck length. The period of the photon's orbit was proportional to the Planck time and the mass energy of each photon is proportional to the Planck mass.

The concept of a Planck length, Planck mass and Planck time all emerge from this single model.

Primal Proofs offer several proofs that deal with prime numbers. A proof by contradiction of Goldbach's binary conjecture that every even natural number greater than two (2) can be expressed as the sum of two (2) primes is given. A proof of Goldbach's ternary conjecture that all natural numbers greater than five (5) are the sum of three (3) primes via the binary proof is presented. A proof by construction (utilizing the proof of the binary conjecture) of the twin prime conjecture is offered. Proofs of the Riemann hypothesis by both deduction and contradiction are presented. A proof that any prime greater than three (3) is the mean of two other primes is presented. A proof is offered that any even number greater than fourteen (14) satisfies Goldbach's binary conjecture in a plurality of ways. Two entangled formulas that generate all the primes beyond the second prime ($P_2 = 3$) are developed and summarized.

Acknowledgments

I acknowledge God for providing all the resources necessary to gain some insight as to the nature of nuclear forces.

I acknowledge my country for providing me with all my freedoms especially, freedom of speech.

I acknowledge my parents for providing a secure and nurturing environment that initially made my learning fun.

I acknowledge all my teachers especially my science and mathematics teachers.

I acknowledge every author in the reference section of this book for providing both ideas and data.

Preface

This study was undertaken to investigate the theoretical nature of the strong nuclear force. Both protons and neutrons are composed of three quarks each of which carry a red, green or blue color charge. Thus, there is a strong nuclear force between the three quarks which exist inside nucleons (protons or neutrons). According to the standard particle model, it is the gluons which are responsible for this strong nuclear force.

There are also forces between the nucleons (protons and neutrons) which compose all atomic nuclei. According to the standard model, it is the pions which are responsible for these forces. The shell model of the nucleus describes how these protons and neutrons form energy shells which exist in all atomic nuclei.

The laws of physics and the standard particle model are also provided. All basic physical units and basic physical constants are included.

Table of Contents

Leading Pages i

Title Page ii

Copyright Page iii

Dedication iv

About the Author v

Other Works By Author vi

Acknowledgments x

Preface xi

Table of Contents xii

Chapter 1 Quark Color 1

Chapter 2 Source of Color 11

Chapter 3 Relativistic Harmonic Oscillator 21

Chapter 4 Proton Quark Dynamics 29

Chapter 5 Color Spring Constant 38

Chapter 6 Wave Mechanical Preliminaries 46

Chapter 7 Proton Wave Equation 55

Chapter 8 Angular Momentum 61

Chapter 9 Light Nuclei 68

Chapter 10 Elementary Particles 79

Chapter 11 Cold Fusion 96

Glossary 105

Figure 1 3

Figure 2 10

Figure 3	10
Figure 4	12
Figure 5	30
Figure 6	71
Figure 7	73
Figure 8	76
Table 1	75
Fundamental Physical Laws	114
Basic Units	142
Basic Physical Constants	143
Basic Elementary Particles	145
References	159

Index 167

Trailing Pages 184 – 185

Chapter 1

Quark Color

Both the proton (which is stable) and the neutron (which if isolated, decays into a proton, electron and anti-neutrino) are composed of 3 quarks. Please refer to the basic elementary particles section towards the end of this book. The quarks not only carry fractional electrical charge, but also carry a new type of quantity called color charge (which has nothing to do with ordinary color). Color charge exists as three types, called red, green or blue. Color charge can also be anti-

2 The Nuclear Force

red, anti-green or anti-blue. This is similar to ordinary electrical charge existing as a plus (+) type or an anti-plus (negative (−)) type.

Color charge may be defined as

(C1) $Q_R = 1/2(1 + i3^{1/2})$

(C2) $Q_G = -1$

(C3) $Q_B = 1/2(1 - i3^{1/2})$

where R, G and B stand for red, green and blue color respectively. i is defined as

(C4) $i = -1^{1/2}$

which occurs in the ordinary definition of complex numbers. Note the following relationships between the color charges.

(C5) $Q_R + Q_G + Q_B = 0$

(C6) $Q_R Q_G = -Q_R$
(C7) $Q_B Q_G = -Q_B$
(C8) $Q_R Q_B = -Q_G = 1$

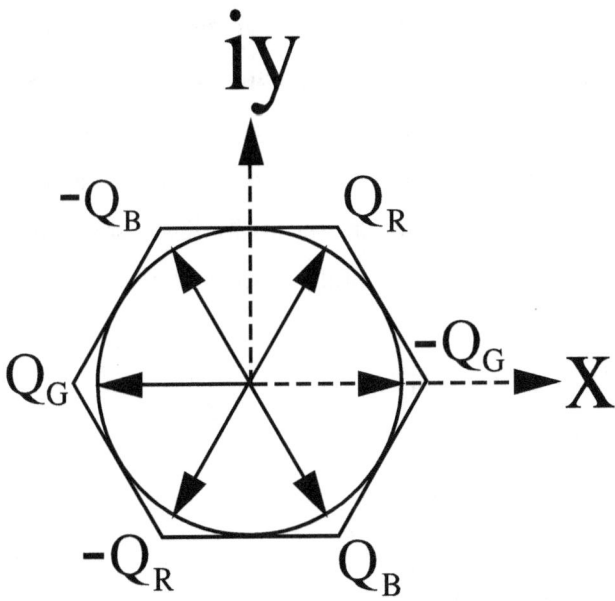

Figure 1 Colors Charges and Anti-Color Charges

4 The Nuclear Force

The quark color charges may be represented by a hexagon circumscribing a unit circle (radius equal to one (1)) in the complex plane. The vertices of the hexagon are at precisely the correct angles of the color charges and anti-color charges. This is all shown in Figure 1 above. Red, green and blue color charges have an angular separation of 120 degrees. Similarly, Anti-red (–R), anti-green (–G) and anti-blue (–B) color charges are all 120 degrees apart. Note that anti-colors are shown by the dashed line arrows inside the unit circle. Also, note that

(C1.1) $Q_R^2 = -Q_B$

(C2.1) $Q_G^2 = -Q_G = 1$

(C3.1) $Q_B^2 = -Q_R$

Protons and Neutrons

The standard model of particles includes trios of up and down quarks which comprise both the neutron and proton. Each quark has axial (spin) angular momentum of $1/2\hbar$ where $\hbar$ is Planck's constant (h) divided by 2π or

(C9) $S = 1/2\hbar = 1/2(h/2\pi)$

where h is Planck's constant and S is axial angular momentum called spin. Planck's constant divided by 2π is known as Planck's rationalized constant.

The proton consists of two (2) up quarks and one (1) down quark. The up quark has an electrical charge of two thirds the charge on a positron (anti-electron) denoted by $2/3\ e^+$. The down quark has an electrical charge of negative one third the charge on a positron and denoted by $-1/3\ e^+$.

The neutron consists of two (2) down quarks and one (1) up quark.

One of the up quarks of a proton is red. The other up quark is blue. The down quark of a proton is green. Any other permutation of colors among the proton's quarks is possible.

One of the down quarks of a neutron is red. The other down quark is blue. The up quark of a neutron is green. Any other permutation of colors among the neutron's quarks is possible.

Figure 2 is a model of the proton. Figure 3 presents a model of the neutron.

The electrical coulomb force of repulsion between the two up quarks in a proton is four times ~$4/9e^{+2}$ that of the electric force of repulsion between the two down quarks in a neutron ~$1/9e^{+2}$. The electrical coulomb force of attraction between the up and down quarks are the same for the neutron as in the proton ~$2/9e^{+2}$.

Pions

The pi minus (π^-) is composed of a colored down quark with a charge of $-1/3$ e^+ and an anti-colored, anti-up quark with a charge of $-2/3$ e^+ for a colorless total charge of e^-. The pi zero (π^0) is a mixture of a colored up quark and an anti-colored, anti-up quark, with a colored down quark and an anti-colored, anti-down quark for a colorless total charge of zero. The pi plus (π^+) is composed of an colored up quark with a charge of +2/3 e^+ and an anti-colored, anti-down quark with a charge of +1/3 e^+ for a colorless total charge of e^+. In summary, these pions have no color and carry charges of $-e^+$, 0 and e^+ respectively. Pions have spin angular momentum of 0. Note that in this notation,

(C9.1) $e^- = -e^+$

Electric Charge and Color

It is well known that macroscopic electrical charge is always an integer (including zero) multiple of the charge on a positron (e^+) or the charge on an electron (e^-).

Note that the positive, negative and neutral electrical charge may be expressed as a linear combination of colors most simply as seen by equations

(C10) $e^+ = e^+(2Q_R + Q_G + 2Q_B)$

8 The Nuclear Force

(C11) $e^- = e^-(2Q_R + Q_G + 2Q_B)$ and

(C12) $0 = (Q_R + Q_G + Q_B)$

This means that the positive charge of any atomic nucleus or the total charge of its electron cloud may be expressed as a linear combination of red, green and blue color. Additionally, the charge on a pion can also be expressed as a linear combination of color and anti-color as expressed by the following equations

(C13) $q_{\pi+} = (1/3)e^+(-Q_C) - (2/3)e^+Q_C$
$\qquad = -e^+Q_C$

(C14) $q_{\pi-} = (2/3)e^+(Q_C) - (1/3)e^+(-Q_C)$
$\qquad = e^+Q_C$

However, in order that the pion charge is real, C must be green which will reduce equations (C13) and (C14) to

$$\text{(C13.1)} \quad q_{\pi+} = (1/3)e^+(-Q_C) - (2/3)e^+Q_C$$
$$= -e^+Q_C = -e^+Q_G = -e^+(-1)$$
$$= e^+$$

$$\text{(C14.1)} \quad q_{\pi-} = (2/3)e^+(Q_C) - (1/3)e^+(-Q_C)$$
$$= e^+Q_C = e^+Q_G = e^+(-1)$$
$$= -e^+$$

$$\text{(C15)} \quad q_{\pi 0} = (1/3)e^+(-Q_C) + (1/3)e^+Q_C$$
$$= 0$$

$$\text{(C16)} \quad q_{\pi 0} = (2/3)e^+(-Q_C) + (2/3)e^+Q_C$$
$$= 0$$

where in equations (C15) or (C16), C can either be red (R) green(G) or Blue (B).

10   The Nuclear Force

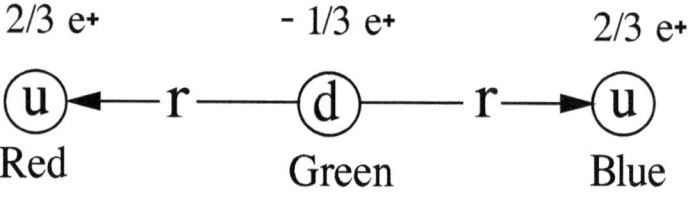

Figure 2 Proton Quarks

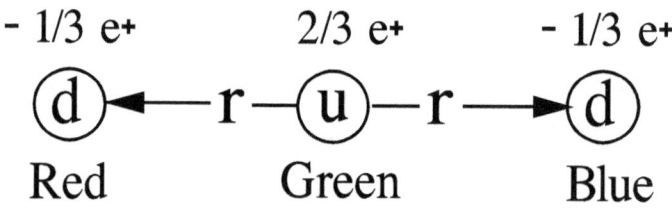

Figure 3 Neutron Quarks

Chapter 2

Source of Color

The source of the color field is color density. This is analogous to the source of the electric field being the charge density. However, each color density is postulated to be constant. This means that by Gauss's law, the sources of all three vector color fields (**R**, **G**, **B**) are their respective constant color densities (ρ_R, ρ_G, ρ_B). V_0 is a constant volume assumed to be the same for any color. Thus,

(S1) $\nabla \cdot \mathbf{R} = \rho_R = Q_R/V_0$

12 The Nuclear Force

(S2) $\nabla \bullet \mathbf{G} = \rho_G = Q_G/V_0$

(S3) $\nabla \bullet \mathbf{B} = \rho_B = Q_B/V_0$

(S4) $\nabla \bullet (\mathbf{R} + \mathbf{G} + \mathbf{B}) = (Q_R + Q_G + Q_B)/V_0 = 0$

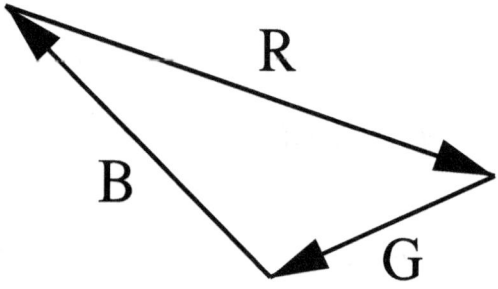

Figure 4   Color Trio Fields Add To Zero

Note that equation (S4) is analogous to the Maxwell's equation $\nabla \bullet \mathbf{B}_M = 0$ which says that the source of a magnetic field, $\mathbf{B}_M$ is zero which also means that magnetic fields always exist in closed loops. Another way of looking at equation (S4) is that when the red, green and blue color (as in figure 4 above) field vectors are added tip to tail, they always form a closed triangle. Moreover, equation (S4) says that there is no source of the sum of all three color fields. Integrating the three equations (S1), (S2) and (S3) yields

(S1.1) $|\mathbf{R}|4\pi r_R^2 = \rho_R \pi r_R^3 4/3$

or

(S1.2) $|\mathbf{R}| = (\rho_R/3) r_R$

where r_R is the radial distance from the red color Q_R. Thus, equation (S1.2) can be written in vector notation as

(S1.3) $\mathbf{R} = (\rho_R/3)\mathbf{r}_R = Q_R \mathbf{r}_R/(3V_0)$

Similarly for the green field

(S2.1) $|\mathbf{G}|4\pi r_G^2 = \rho_G \pi r_G^3 4/3$

14 The Nuclear Force

or

(S2.2) $|\mathbf{G}| = (\rho_G/3)r_G$

In vector notation becomes

(S2.3) $\mathbf{G} = (\rho_G/3)\mathbf{r}_G = Q_G\mathbf{r}_G/(3V_0)$

Similarly for the blue field

(S3.1) $|\mathbf{B}|4\pi r_B^2 = \rho_B \pi r_B^3 4/3$

or

(S3.2) $|\mathbf{B}| = (\rho_B/3)r_B$

and in vector form is,

(S3.3) $\mathbf{B} = (\rho_B/3)\mathbf{r}_B = Q_B\mathbf{r}_B/(3V_0)$

Note that the magnitude of all the color fields increase with the distance, r from their color sources.

Color Forces

The force on an electrical charge q, is given by the Lorentz force as:

(CF0) $\mathbf{F}_q = q[\mathbf{E} + \mathbf{v}_q \times \mathbf{B}_M]$

where $\mathbf{F}_q$ is the force on a charge, q in an external electric field, $\mathbf{E}$ and external magnetic field, $\mathbf{B}_M$. Here $\mathbf{v}_q$ is the velocity of the charge q. For a static charge, its velocity is zero and equation (CF0) reduces to

(CF0.1) $\mathbf{F}_q = q\mathbf{E}$

Analogously, the force on a static color is presumed be of the same form as in equation (CF0.1). Hence, the force on a red color, Q_R in the presence of a stationary green and blue color field is:

16 The Nuclear Force

(CF1.0) $\mathbf{F}_R = Q_R[\mathbf{G} + \mathbf{B}]$

and because of (S2.2) and (S3.2) becomes

(CF1.1) $\mathbf{F}_R = Q_R[Q_G\mathbf{r}_G + Q_B\mathbf{r}_B]/(3V_0)$

and which by equations (C6) and (C8) reduces to

(CF1.2) $\mathbf{F}_R = -[Q_R\mathbf{r}_G + Q_G\mathbf{r}_B]/(3V_0)$
$= -\rho_R\mathbf{r}_G/3 - \rho_G\mathbf{r}_B/3$
$= -k_R\mathbf{r}_G - k_G\mathbf{r}_B$

where $k_R = \rho_R/3$ and $k_G = \rho_G/3$. Similarly, the force on a green color, Q_G in the presence of a stationary red and blue color field is:

(CF2.0) $\mathbf{F}_G = Q_G[\mathbf{R} + \mathbf{B}]$

reducing to

(CF2.1) $\mathbf{F}_G = Q_G[Q_R\mathbf{r}_R + Q_B\mathbf{r}_B]/(3V_0)$

and which by equations (C6) and (C7) reduces to

(CF2.2) $\mathbf{F}_G = -[Q_R\mathbf{r}_R + Q_B\mathbf{r}_B]/(3V_0)$

$\qquad = -\rho_R\mathbf{r}_R/3 - \rho_B\mathbf{r}_B/3$

$\qquad = -k_R\mathbf{r}_R - k_B\mathbf{r}_B$

where $k_B = \rho_B/3$. Similarly, the force on a blue color, Q_B in the presence of a stationary red and green color field is:

(CF3.0) $\mathbf{F}_B = Q_B[\mathbf{R} + \mathbf{G}]$

reducing to

(CF3.1) $\mathbf{F}_B = Q_B[Q_R\mathbf{r}_R + Q_G\mathbf{r}_G]/(3V_0)$

and which by equations (C7) and (C8) reduces to

(CF3.2) $\mathbf{F}_B = -[Q_G\mathbf{r}_R + Q_B\mathbf{r}_G]/(3V_0)$

$\qquad = -\rho_G\mathbf{r}_R/3 - \rho_B\mathbf{r}_G/3$

$\qquad = -k_G\mathbf{r}_R - k_B\mathbf{r}_G$

Color Confinement

Equation (CF1.2) describe the force on a red quark due to the presence of a blue and green quark. Equation (CF2.2) describe the force on a green quark due to red and blue quark. Equation (CF3.2) describe the force on a blue quark due to red and green quark.

Note that the color force on any one quark is proportional to the distance it is from the other two quarks. Thus, if the distances of any quark from the other two is increased, then the force is also increased. Since the potential energy is proportional to this binding force, it also must increase. Thus, rather than removing and isolating a single quark, this potential energy dissipates by converting to other available forms of energy (particles). This is the reason that quark trios are bound to each other and it is the reason that it is experimentally found that quarks cannot be isolated from each other.

Note also that the quarks in a quark trio are bound to one another by analogous springs as described by Hooke's law which says that

(CF4.0) $F = -kr$

where k is the spring constant and r describes either the expansion or compression of the spring. Recall that the solution to equation (CF4.0) describes a simple harmonic oscillator. Thus, a proton may be viewed as three quarks attached to each other by color springs and damped by the electrical coulomb forces between their charges.

The potential energy, V(r) associated with the force of equation (F4.0) is

(CF4.1) $V(r) = (k/2)r^2$

Even though the force between quarks is directly proportional to the distance of their separation, it is experimentally known that the direction of this force is also dependent on the angular momentum states of the quarks. These are the so called spin-

spin coupling and the spin-orbit coupling forces associated between neutrons and protons. Since neutrons and protons are composed of quarks, then there must be some rules for calculating the direction of the strong Hooke color forces between quarks. Let us now describe a relativistic harmonic oscillator. This is relevant since Hooke forces produce oscillator type motion.

Chapter 3

Relativistic Harmonic Oscillator

In order to integrate the force equation for a relativistic harmonic oscillator, a form for the relativistic definition of Newtonian force will be developed. The starting point for this is the well known relativistic relationship between a mass m, moving with velocity v given by

(R1.0) $m = m_0(1-(v/c)^2)^{-1/2}$

where m_0 is the rest mass (mass when it is not moving) and c is the velocity of light in free space. Taking a time derivative of equation (R1.0) yields

(R1.1) $dm/dt = (mv/(c^2 - v^2))dv/dt$

which means linear relativistic acceleration is

(R1.2) $dv/dt = ((c^2 - v^2)/mv)dm/dt$

and also implys that

(R1.2.1) $dv/dt \rightarrow 0$ in the limit as $v \rightarrow c$

The definition of relativistic force is

(R3.0) $F = dmv/dt = mdv/dt + vdm/dt$

and plugging in from equation (R1.2) results in

(R3.1) $F = m((c^2 - v^2)/mv)dm/dt + vdm/dt$

and reduces the relativistic force to

(R3.2) $F = (c^2/v)dm/dt$

where v is assumed to be in the same direction as the force F. Hence, In vector notation (bold characters) equation (R3.2) may be expressed as

(R3.3) $\mathbf{F} = (c^2/v^2)(dm/dt)\mathbf{v}$

Derivation of E=mc²

As a check, note that the classic definition of kinetic energy, T is the integral of force, **F** through a vector displacement, **r** written as

(R3.2.1) $T = \int \mathbf{F} \bullet d\mathbf{r}$

where **F** and **r** are vectors and the dot (•) indicates the normal scalar product of two vectors. If (R3.3) is plugged into (R3.2.1) the result is

(R3.2.2) $T = \int F dr = \int [(c^2/v^2) dm/dt] \mathbf{v} \bullet \mathbf{dr}$

but since $\mathbf{v} = \mathbf{dr}/dt$, (R3.2.2) reduces to

(R3.2.3) $T = c^2 \int [(\mathbf{v} \bullet \mathbf{v})/(v^2)] dm = c^2 \int dm$

which upon integration yields

(R3.2.4) $T = mc^2 - m_0c^2 = (m - m_0)c^2$

Note that (m_0c^2) is the negative integration constant and interpreted as rest mass energy (i.e. if m is at rest then $m = m_0$ and $T = 0$)

If the rest mass energy is interpreted to be potential energy $V = m_0c^2$, the total energy, E_T is defined to be the kinetic energy, T plus the potential energy, V thus,

(R3.2.5) $E_T = T + V = mc^2 - m_0c^2 + m_0c^2 = mc^2$

to formally yield Einstein's famous result. This means that the tools of equations

(R3.2) $F = (c^2/v)dm/dt$ and

(R3.3) $\mathbf{F} = (c^2/v^2)(dm/dt)\mathbf{v}$

is consistent with special relativity. Together equations (R3.2.4) and (R3.2.1) results in

(R4.0) $T = \int \mathbf{F} \bullet d\mathbf{r} = c^2 \int dm$

Together with

(R3.2.4) $T = mc^2 - m_0c^2 = (m - m_0)c^2$

equation (R4.0) reduces to

(R5.0) $mc^2 - m_0c^2 = \int \mathbf{F} \bullet d\mathbf{r}$

For a relativistic harmonic oscillator, the force on a mass at the end of a strong spring is anti-parallel to the displacement vector **r**. Note that at the center of the mass which has not been displaced is the origin (zero point) of the displacement vector **r**. The force is thus,

(R6.0) $\mathbf{F} = -k\mathbf{r}$

and

(R6.1) $mc^2 - m_0c^2 = -k\int_{r_0}^{r} r\, dr$

where $|\mathbf{r}| = r$

When $r = r_0$, the end point of the oscillation, the mass of the oscillator has no velocity. Therefore, the oscillator's mass is the rest mass at the end point. Nevertheless, equation (R6.1) reduces to

(R6.2) $mc^2 - m_0c^2 = k/2(r_0^2 - r^2)$

which can be rewritten as

(R6.2) $mc^2 - m_0c^2 + kr^2/2 = kr_0^2/2$

which may be considered an expression for the total energy as

(R6.3) $E_T = T + V = kr_0^2/2 =$ constant

where $T = mc^2 - m_0c^2$ is the kinetic energy and

(R6.4) $V = kr^2/2$

is the potential energy. Note that this harmonic oscillator potential energy is in between a square well potential ($kr_0^2/2$) and the so-called oscillator potential energy ($kr^2/2 - kr_0^2/2$) described in the shell model of the nucleus of atoms. Thus, the color forces which were presented in chapter two offer a basis for the shell model of the nucleus of all atoms.

Note that in the total energy equation of (R6.2), when $r = \pm r_0$, then $m = m_0$ which correctly predicts that a vibrating mass is at rest (while changing direction) at either end point of oscillation.

Chapter 4

Proton Quark Dynamics

What follows offers an analysis of proton dynamics based on the model of figure 5 below. Choose a coordinate system where the vertical axis (y) passes through the center of the down quark in a proton. The horizontal (x) axis is chosen to go through all three quark centers. Quark spin, quark charge, quark species and quark color are also listed. Consider the forces on the far right up quark in figure 5. There are electrical coulomb

30   The Nuclear Force

forces as well as color forces which act upon this up quark. The electrical force of attraction between the up quark electrical charge and the down quark electrical charge is

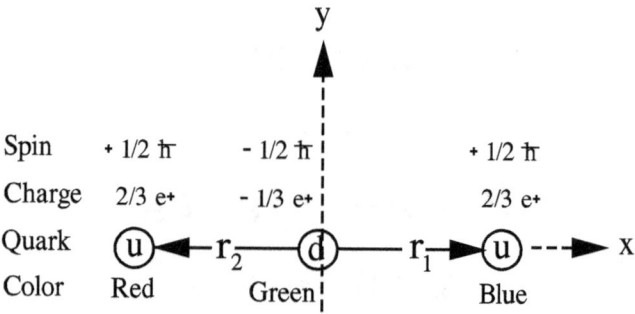

Figure 5   Proton Coordinate System

(F1.1) $F_{e+} = K_C(2/3)(-1/3)e^2/r^2 = -K_C(2/9)e^2/r^2$

where K_C is Coulomb's constant and e is the charge on a positron. This force on the right up quark points in the negative x direction. The radii of both up quarks are r_1 and r_2 respectively. It will be assumed due to the symmetry that

(F2.1) $r = |\vec{r}_1| = |\vec{r}_2|$

as shown in Figure 5. The electrical force of repulsion between the two up quarks is

(F1.2) $F_{e-} = K_C(2/3)(2/3)e^2/(2r)^2 = K_C(4/9)e^2/(2r)^2$

which points in the positive x direction. Thus, the total electrical force acting on the right (Figure 5) up quark is

(F1.3) $F_{eT} = -K_C(2/9)e^2/r^2 + K_C(4/9)e^2/(2r)^2$
$= -K_C(1/9)e^2/r^2$

This force points in the negative x direction. The color force acting on the right up quark (Figure 5) due to the down quark is given by

(F3.1) $F_{Cud} = -kr$

where k is the color spring constant and again r is given by equation (F2.1) as depicted in Figure 5.

32 The Nuclear Force

This assumption will be later justified when spin-spin and spin-orbit coupling forces are described. It will also be assumed that the restoring color force acting on the right up quark due to the other up quark is given by

(F3.2) $F_{Cuu} = 2kr$

The total color force on the right up quark is the sum of these color forces and is given by

(F3.3) $F_{CT} = -kr + 2kr = kr$

Thus, the sum total force acting on the right up quark is the sum of the total electric Coulomb force plus the total color force. This is given by

(F4.1) $F_T = F_{eT} + F_{CT}$
$= -K_C(1/9)e^2/r^2 + kr$

However for the up quark, the total force F_T must be zero at the minimal end point when $r = r_0$. Thus,

(F4.2) $kr_0^3 = K_C(1/9)e^2$

where r_0 is assumed to be the oscillatory minimal end point radial distance between the up quark and the central down quark (see figure 5).

Total Energy of the Proton

As a check, consider the total energy of the proton, E_{PT}. Since the coordinate system is sitting on the down quark, the down quark has total energy equal to its rest mass energy, $m_{d0}c^2$. Each of the two up quarks total energy is $m_u c^2$. The total electrical potential energy consists of three terms, namely the electrical potential energy between the two up quarks, V_{euu}, the electrical potential energy between one up quark and the down quark, V_{eud} and the electrical potential energy between the other up quark and the down quark, V_{eud}.

Similarly, the total color potential energy also consists of three terms, namely the color potential

34 The Nuclear Force

energy between the two up quarks, V_{Cuu}, the color potential energy between one of the up quarks and the down quark, V_{Cud} and the color potential energy between the other up quark and the down quark, V_{Cud}. Thus, we have for the total energy of the proton

$$(E1.1) \quad E_{PT} = m_{d0}c^2 + 2(m_u c^2) + V_{euu} + 2(V_{eud})$$
$$+ V_{Cuu} + 2(V_{Cud})$$
$$= \text{constant}$$

The repulsive Coulomb potential energy between the two up quarks, namely V_{euu} is

$$(E1.2) \quad V_{euu} = -K_C(4/9)e^2/2r$$

The attractive (negative) Coulomb potential energy between one of the up quarks and the down quark, namely, V_{eud} is

$$(E1.3) \quad V_{eud} = K_C(2/9)e^2/r$$

Thus, the total Coulomb potential energy is

(E1.4) $V_{eT} = V_{euu} + 2(V_{eud})$
$= -K_C(4/9)e^2/2r + 2(K_C(2/9)e^2/r)$
$= (2/9)K_C e^2/r$

The color potential energy between the two up quarks, V_{cuu} is

(E1.5) $V_{Cuu} = 1/2k(2r)^2$

The color potential energy between the up quark and the down quark is

(E1.6) $V_{Cud} = -1/2kr^2$

Thus, the total color potential energy is

(E1.7) $V_{CT} = V_{Cuu} + 2(V_{Cud})$
$= 1/2k(2r)^2 - 2(1/2kr^2)$
$= kr^2$

36 The Nuclear Force

Plugging equations (E1.2) to (E1.7) into equation (E1.1) yields

(E2.1) $E_{PT} = m_{d0}c^2 + 2(m_u c^2) + (2/9)K_C e^2/r + kr^2$

Since the total proton energy is a constant, its derivative with respect to r must vanish. Thus,

(E3.1) $dE_{PT}/dr = 0$

This implies for that at end point $r = r_0$

(E3.2) $-(2/9)K_C e^2/r_0^2 + 2kr_0 = 0$

since the masses of the quarks are assumed independent of r making

(E2.2) $dm_{d0}/dr = dm_u/dr = 0$

Solving equation (E3.2) for kr_0^3, yields

(E3.3) $kr_0^3 = K_C(1/9)e^2$

which is precisely the same as equation

(F4.2) $kr_0^3 = K_C(1/9)e^2$

Note that equation (F4.2) was achieved by considering the force on one of the up quarks, while equation (E3.3) was derived by considering the total energy of a proton. Note that the color force obeys Hooke's law. It is this force that represents the behavior of the gluons which are responsible for the strong nuclear force of the standard model of particle physics.

Chapter 5

Color Spring Constant

Consider the total force (both electrical and color) on the proton's up quark was previously presented as equation

(F4.1) and renumbered for this chapter is given by

(CS1) $F_T = F_{eT} + F_{CT} = -K_C(1/9)e^2/r^2 + kr$

where K_C is coulombs constant, e is the charge on an electron and k is the color spring constant. Recall that the kinetic energy of a particle may be computed if the forces on this particle are known. This was given by equation (R5.0) and renumbered for this section as

(CS2) $mc^2 - m_0c^2 = \int \mathbf{F} \bullet d\mathbf{r}$

If the forces of equation (CS1) on the up quark of a proton is plugged into equation (CS2) the result is

(CS3) $mc^2 - m_0c^2 = \int\limits_{r}^{r_0} (-K_C(1/9)e^2/r^2 + kr\,)dr$

Performing the integration, the result is

(CS4) $mc^2 - m_0c^2 = -K_C(1/9)e^2/r + K_C(1/9)e^2/r_0 + kr_0^2/2 - kr^2/2$

Note that when $r = r_0$, the relativistic kinetic energy, $mc^2 - m_0c^2$ is zero. Rewriting equation (CS4)

40 The Nuclear Force

(CS5) $mc^2 - m_0c^2 + K_C(1/9)e^2/r + kr^2/2 =$
$K_C(1/9)e^2/r_0 + kr_0^2/2 = E_{Tup}$

which is the total energy of one up quark in a proton. Thus the total energy of a proton must be twice the energy E_{Tup} plus the rest mass energy of the proton's down quark. This means

(CS6) $m_P c^2 = K_C(2/9)e^2/r_0 + kr_0^2 + m_{d0}c^2$

where m_P is the mass of the proton and m_{d0} is the rest mass of the down quark. Since $m_{d0} \ll m_P$ and because of equation (F4.2) $kr_0^3 = K_C(1/9)e^2$ equation (CS6) reduces to

(CS7) $m_P c^2 = 1/3(K_C e^2/r_0)$

Solving equation (CS7) for r_0 results in

(CS8) $r_0 = 1/3(K_C e^2/m_P c^2)$

Recall that coulombs constant is defined to be

(CS9) $K_C = 1/(4\pi\varepsilon_0)$

where ε_0 is the permittivity constant of free space. It is found that the value of coulombs constant is

(CS10) $K_C = 8.99 \times 10^9$ nt-m^2/coul^2

The charge on the electron is measured to be

(CS11) $e = -1.6022 \times 10^{-19}$ coulombs

The mass of the proton is

(CS12) $m_P = 1.67239 \times 10^{-27}$ kilograms

The velocity of light c is

(CS13) $c = 3 \times 10^8$ meters/second

Plugging equations (CS10), (CS11), (CS12) and (CS13) into equation (CS8) results in

(CS14) $r_0 = 5.1 \times 10^{-19}$ meters

Solving for k in equation (F4.2) $kr_0^3 = K_C(1/9)e^2$ one gets

(CS15) $k_0 = K_C(1/9)e^2/r_0^3$

Plugging equations (CS10), (CS11), (CS12), (CS13) and (CS14) into equation (CS15) results in

(CS16) $k_0 = 1.93 \times 10^{29}$ newtons/meter

which establishes the enormity of the color spring constant, k_0.

However considering equation

(CS1) $F_T = F_{eT} + F_{CT} = -K_C(1/9)e^2/r^2 + kr$

and setting the total force on an up quark to zero, we obtain

(CS1.1) $k_p r_p^3 = K_C (1/9) e^2$

Experimentally, however the radius of a proton is

(CS1.2) $r_p = .6 \times 10^{-15}$ meters

Plugging r_p from equation (CS1.2), K_C, and e into equation (CS1.1) yields

(CS1.3) $k_p = 1.19 \times 10^{17}$ newtons/meter

which is 10^{12} times weaker than the spring constant of equation

(CS16) $k_0 = 1.93 \times 10^{29}$ newtons/meter

This can only be true if k is a function of r between $r_0 = 5.1 \times 10^{-19}$ meters and $r_p = .6 \times 10^{-15}$ meters. It will be assumed the k varies linearly with r. Thus,

(CS17) $k = mr + b$

where m and b are constants. Plugging in r_P and r_0, one optains

(CS17.1) $k_0 = mr_0 + b$

and

(CS17.2) $k_P = mr_p + b$

Solving for m and b produces equation

(CS18) $k = (k_0 - k_P)/(r_0 - r_p)r + k_0 - (k_0 - k_P)/(r_0 - r_p)r_0$

where r varies between r_0 and r_p. Thus, it is found that the spring constant is not a constant but varies during the oscillation of the up quarks to produce a consistent value of the nuclear force.

Chapter 6

Wave Mechanical Preliminaries

The following relativistic equation

(W1) $-\hbar^2 \nabla^2 \Psi(\check{r},t)/(m+m_0) + V(r)\Psi(\check{r},t) = E_T \Psi(\check{r},t)$

was derived in a book entitled "An Alternative Lorentz Invariant Relativistic Wave Equation" by the author.

Meaning of The Alternative Relativistic Wave Equation

This equation is Lorentz invariant and properly describes the wave nature of systems including those having zero rest mass, m_0 as well as relativistic mass m. Prior to this equation, there was no way to describe a relativistic quantum mechanical system in which zero rest mass particles move under the influence of a central field of force derivable from a potential V(r). The following discussion will hopefully pave the way for a quantum mechanical description of a proton in terms of its constituent quarks. In equation

(W1) $-\hbar^2 \nabla^2 \Psi(\check{r},t)/(m+m_0) + V(r)\Psi(\check{r},t) = E_T \Psi(\check{r},t)$

m_0 is the rest mass and m is its mass in motion assumed to be a function of its velocity according to

(W1.1) $m = m_0(1 - (v/c)^2)^{-1/2}$

featured in Einstein's special theory of relativity.

48 The Nuclear Force

Referring to equation (W1), the system can be moving under the influence of a central force derivable from a potential V(r). The wave function, $\Psi(\v{r},t)$ is assumed to be a function of position, r (x,y,z) and time, t. When describing stationary states with total energy, E_T, the wave function is assumed to be a function of position only. Again, Planck's constant divided by 2π is denoted by $\hbar$. The Del operator is denoted by

(W2) $\nabla = (\partial/\partial x, \partial/\partial y, \partial/\partial z)$

and Laplacian

(W3) $\nabla^2 = \nabla \cdot \nabla = (\partial^2/\partial x^2, \partial^2/\partial y^2, \partial^2/\partial z^2)$

where • denotes the normal vector dot product. Equation

(W1) $-\hbar^2 \nabla^2 \Psi(\v{r},t)/(m+m_0) + V(r)\Psi(\v{r},t) = E_T \Psi(\v{r},t)$

may be rewritten as

(W4) $\hat{H}\Psi(\v{r},t) = E_T\Psi(\v{r},t)$

where $\hat{H}$ is the relativistic Hamiltonian operator given by

(W5) $\hat{H} = -\hbar^2\nabla^2/(m + m_0) + V(r)$

Hamiltonian Operator For A System of Two Particles

Beginning on page 173 of David Park's book entitled "Introduction to the Quantum Theory" (See reference section) a general two particle wave equation is developed. Following this philosophy, the resulting Hamiltonian operator is

(W6) $\hat{H}_T = \v{T}_1 + \v{T}_2 + V(\v{r}_1 - \v{r}_2)$

where $V(\v{r}_1 - \v{r}_2)$ is the potential energy between the two particles at vector positions $\v{r}_1$ and $\v{r}_2$. $\v{T}_1$ and $\v{T}_2$

are the kinetic energy operators for both particles and $\hat{H}_T$ is the Hamiltonian operator for the combined system. This implies

(W7) $\hat{H}_T \Psi_T(\check{r}_1,\check{r}_2) = E_T \Psi_T(\check{r}_1,\check{r}_2)$

where E_T is the total energy eigenvalue for the possible stationary energy states of both particles. The wave function representing both particles is $\Psi_T(r_1,r_2)$ and it is assumed that

(W8) $\Psi_T(\check{r}_1,\check{r}_2) = \Psi(\check{r}_1)\Psi(\check{r}_2)$

Kinetic Energy Operators

Furthermore, by the form of equation

(W1) $-\hbar^2 \nabla^2 \Psi(\check{r},t)/(m+m_0) + V(r)\Psi(\check{r},t) = E_T \Psi(\check{r},t)$

it is assumed that the kinetic energy operators obey equations

(W9) $\check{T}_1 = -\hbar^2 \nabla_1^2 / (m_1 + m_{10})$ and

(W10) $\check{T}_2 = -\hbar^2 \nabla_2^2 / (m_2 + m_{20})$

where m_1, m_{10}, m_2 and m_{20} are the masses and rest masses of the two systems respectively. Moreover, the two Laplacians of equations (5.9) and (5.10) are defined by

(W11) $\nabla_1^2 = (\partial^2/\partial x_1^2, \partial^2/\partial y_1^2, \partial^2/\partial z_1^2)$ and

(W12) $\nabla_2^2 = (\partial^2/\partial x_2^2, \partial^2/\partial y_2^2, \partial^2/\partial z_2^2)$

where the position (see Figure 5) of particle one is

(W13) $r_1 = (x_1, y_1, z_1)$ and

simultaneously the position of particle two is

(W14) $r_2 = (x_2, y_2, z_2) = (-x_1, -y_1, -z_1) = -r_1$

Alternate Form of Laplacian Operator

On pages 516 – 518 of David Park's book entitled "Introduction to the Quantum Theory" (See reference section) it can be deduced that the Laplacian operator is

(W15) $\nabla^2 = d^2/dr^2 + (2/r)d/dr - \hat{L}^2/r^2$

where

(W15.1) $\hat{L}^2 = -[(1/(\sin\theta))(\partial/\partial\theta(\sin\theta\partial/\partial\theta) + (1/\sin^2\theta)(\partial^2/\partial\varphi^2)]$

$\hat{L}^2$ is the square of the orbital angular momentum operator, θ is the angle that the projection of the radial distance, r onto the xy plane, makes with the x axis and φ is the angle the radial distance, r makes with the z axis (normal spherical coordinates). It turns out that the eigenvalues of $\hat{L}^2$ performed on a normalized spherical wave function described by r, θ and φ, $\Psi(r,\theta,\varphi)$ is

(W15.2) $\hat{L}^2 \Psi(r,\theta,\varphi) = \ell(\ell+1)\Psi(r,\theta,\varphi)$

where the orbital angular momentum quantum number, ℓ can only take on non–negative integer values (0, 1, 2, ..). If the wave function depends only on r or if the system is in its lowest angular momentum state ($\ell = 0$), then equation (W15.2) reduces to

(W15.3) $\hat{L}^2 \Psi(r) = 0$

This completes the development of the preliminary tools, which will be used to describe a system of one up quark in a proton (consisting of two up quarks and one down quark). The up quark is influenced by a potential energy function, V(r) describing both a color field and electrical field. Using relativistic wave mechanics, if one up quark can be described by a wave function, then by symmetry, the other can also be described. If the coordinate system is placed on the down quark, its wave function collapses since it doesn't move. Thus, it is hoped that the entire wave function of a proton can be obtained. Since a neutron can be considered an excited state of a proton, it is also anticipated that a neutron's complete wave function may be similarly obtained.

Chapter 7

Proton Wave Equation

In chapter 4, a formula was developed which gives the total energy of a proton as

(E2.1) $E_{PT} = m_{d0}c^2 + 2(m_u c^2) + (2/9)K_C e^2/r + kr^2$

Since the coordinate system of Figure 5 is located in the center of mass of both up quarks, the total potential energy associated with a proton is just the last two terms of equation (E2.1). This is given by equation

(P1.0) $V(r) = (2/9)K_C e^2/r + kr^2$

This equation represents the total potential energy of one up quark in a proton. It can be thought of as the sum of the total Coulomb potential energy plus the total color potential energy. Recall a two particle wave equation developed in chapter 6 is described in general by

(W6) $\hat{H}_T = \check{T}_1 + \check{T}_2 + V(\check{r}_1 - \check{r}_2)$

where $V(\check{r}_1 - \check{r}_2)$ is the potential energy between the two particles at vector positions $\check{r}_1$ and $\check{r}_2$. $\check{T}_1$ and $\check{T}_2$ are the kinetic energy operators for both particles and $\hat{H}_T$ is the Hamiltonian operator for the combined system. This implies

(W7) $\hat{H}_T \Psi_T(\check{r}_1,\check{r}_2) = E_T \Psi_T(\check{r}_1,\check{r}_2)$

where E_T is the total energy eigenvalue for the possible stationary energy states of both particles. The wave function representing both particles is $\Psi_T(\check{r}_1,\check{r}_2)$ and it is assumed that

(W8) $\Psi_T(\check{r}_1,\check{r}_2) = \Psi(\check{r}_1)\Psi(\check{r}_2)$

Thus, the wave equation for both up quarks in figure 5 is

(P1.2) $-\hbar^2\nabla_1^2\Psi_T/(m_U+m_{U0}) - \hbar^2\nabla_2^2\Psi_T/(m_U+m_{U0})$
$\qquad + [(2/9)K_Ce^2/r + kr^2]\Psi_T = E_T\Psi_T$

where

(P1.3) $\check{T}_1 = -\hbar^2\nabla_1^2\Psi_T/(m_U+m_{U0})$

and

(P1.4) $\check{T}_2 = -\hbar^2\nabla_2^2\Psi_T/(m_U+m_{U0})$

and $\Psi_T = \Psi_T(\check{r}_1, \check{r}_2)$ is the wave function of both up quarks. The mass and rest mass of the up quark is m_U and m_{U0} respectively. The energy eigenstate of both up quarks is given by E_T. All other terms are labeled and pictorially represented by figure 5. Since $\check{r}_2 = -\check{r}_1$ implies $\nabla_2 = -\nabla_1$ means that

(P1.5) $\nabla_1^2 = \nabla_2^2 = \nabla^2$

so that equation (P1.2) simplifies to

(P2) $-2\hbar^2 \nabla^2 \Psi_T / (m_U + m_{U0}) +$
$[(2/9)K_C e^2/r + kr^2]\Psi_T = E_T \Psi_T$

If the wave function, Ψ_T for both up quarks in equation (P2) can be solved, then the combined wave function for both up quarks can be described. Moreover, since the wave function for the down quark is collapsed (since by figure 5, it is stationary), it will be represented by a constant. In turn, this means that the entire wave function for a proton may be determined.

Equation (P2) can be put into an alternate form by utilizing equation

(W15) $\nabla^2 = d^2/dr^2 + (2/r)d/dr - \hat{L}^2/r^2$

of the previous chapter. Multiplying equation (P2) by the quantity $(m_U+m_{U0})/(-2\hbar^2)$ yields

(P2.1) $\nabla^2 \Psi(\check{r}) -$
$\quad [(m_U+m_{U0})/\hbar^2][(1/9)K_C e^2/r + (1/2)kr^2]\Psi(\check{r})$
$\quad + [(m_U+m_{U0})/(2\hbar^2)]E_T \Psi(\check{r}) = 0$

Plugging equation (W15) into equation (P2.1) results in

(P3) $[d^2/dr^2 + (2/r)d/dr - \hat{L}^2/r^2]\Psi(\check{r}) -$
$\quad [(m_U+m_{U0})/\hbar^2][(1/9)K_C e^2/r + (1/2)kr^2]\Psi(\check{r})$
$\quad + [(m_U+m_{U0})/(2\hbar^2)]E_T \Psi(\check{r}) = 0$

60 The Nuclear Force

Since by equation

(W15.2) $\acute{L}^2\Psi(\check{r}) = \ell(\ell+1)\Psi(\check{r})$

then equation (P3) can be written as

(P3.1) $[d^2/dr^2 + (2/r)d/dr - \ell(\ell+1)/r^2]\Psi(\check{r}) -$
$[(m_U+m_{U0})/\hbar^2][(1/9)K_Ce^2/r + (1/2)kr^2]\Psi(\check{r})$
$+ [(m_U+m_{U0})/(2\hbar^2)]E_T\Psi(\check{r}) = 0$

The ground state of both up quarks implies that the orbital angular momentum quantum number $\ell = 0$ which reduces equation (P3.1) to

(P4) $[d^2/dr^2 + (2/r)d/dr]\Psi(\check{r}) -$
$[(m_U+m_{U0})/\hbar^2][(1/9)K_Ce^2/r + (1/2)kr^2]\Psi(\check{r}) +$
$[(m_U+m_{U0})/(2\hbar^2)]E_T\Psi(\check{r}) = 0$

It is expected that since wave functions for 1/r potentials and wave functions for r^2 potentials are known, wave functions for the sum of these potentials should be a linear combinations of these.

Chapter 8

Angular Momentum

It is known that in order for the shell model of the nucleus to be successful, the force between nucleons (neutrons or protons) must be dependent on the angular momentum states (spin orbit coupling) of these nucleons. However, nucleons may possess two different forms of angular momentum. These are the orbital angular momentum and the axial or spin angular momentum. Orbital angular momentum can only take on integer (0, 1, 2, ...) values of the rationalized Planck's constant ($\hbar$). Spin angular momentum of nucleons or quarks (which are fermions) have half integer values of the rationalized Planck's constant ($\hbar/2$).

Spin-Spin Coupling Forces

Consider two identical fermions having spin angular momentum, S_1 and S_2 of $\hbar/2$. Form the square of the sum of their spins as

(A1) $(S_1 + S_2)^2 = (S_1 + S_2) \bullet (S_1 + S_2) =$
$S_1 \bullet S_1 + 2 S_1 \bullet S_2 + S_2 \bullet S_2 =$
$S_1^2 + 2 S_1 \bullet S_2 + S_2^2$

where $\bullet$ means the ordinary dot product of two vectors. The sum of two spins $(S_1 + S_2)$ can either add to one (1) when they are parallel (↑↑) or zero (0) when they are anti-parallel (↑↓). The component of one spin in the direction of the other spin is given by the quantity $S_1 \bullet S_2$. Solving equation (A1) for this quantity yields equation

(A2) $S_1 \bullet S_2 = \frac{1}{2}[(S_1 + S_2)^2 - (S_1^2 + S_2^2)]$

For the parallel (↑↑) spins case, equation (A2) reduces to

(A2.1) $S_1 \bullet S_2 (\uparrow\uparrow) = \frac{1}{2}[(1)(1+1) - 2(1/2)(1/2+1)]\hbar^2$
$= (1 - \frac{3}{4})\hbar^2 = \frac{1}{4}\hbar^2$

For the anti-parallel (↑↓) spins case, equation (A2) reduces to

(A2.2) $S_1 \bullet S_2 (\uparrow\downarrow) = 1/2[(0)(0+1) - (1/2)(1/2+1)]\hbar^2$
$= -\frac{3}{4}\hbar^2$

Note that the quantity $S_1 \bullet S_2$ changes sign in going from the parallel case to the anti-parallel case. In fact their relationship is expressed by equation

(A3) $S_1 \bullet S_2 (\uparrow\downarrow) = -3 S_1 \bullet S_2 (\uparrow\uparrow)$

Thus, if the force between identical quarks is dependent on the quantity $S_1 \bullet S_2$, then assume in the case of anti-parallel (↑↓) spins, the force is positive (attractive). Thus, in the case of parallel spins, the force is assumed negative (repulsive). The antiparallel spin force is three times the parallel spin force. These two forces will be referred to as the spin-spin coupling forces.

Spin-Orbit Coupling Forces

Consider two identical fermions having spin angular momentum S_1 and S_2 of $\hbar/2$. In addition, assume they have orbital angular momentum, L_1 and L_2. Form the square of the sum of one spin and the orbital angular momentum of the other as

$$(A4)\ (S_1 + L_2)^2 = (S_1 + L_2)\bullet(S_1 + L_2) =$$
$$= S_1 \bullet S_1 + 2 S_1 \bullet L_2 + L_2 \bullet L_2$$
$$= S_1^2 + 2 S_1 \bullet L_2 + L_2^2$$

Solving for $S_1 \bullet L_2$ yields equation

$$(A4.1)\ S_1 \bullet L_2 = \tfrac{1}{2}[(S_1 + L_2)^2 - (S_1^2 + L_2^2)]$$

For the parallel (↑↑) orbital spin case, equation (A4.1) reduces to

$$(A4.2)\ S_1 \bullet L_2\ (\uparrow\uparrow) = \tfrac{1}{2}[(\ell_2 + \tfrac{1}{2})(\ell_2 + \tfrac{1}{2} + 1) -$$
$$(\tfrac{1}{2})(\tfrac{1}{2}+1) + \ell_2(\ell_2+1)]\hbar^2$$
$$= \tfrac{1}{2}\,[\ell_2^2 + \tfrac{1}{2}\ell_2 + \ell_2 + \tfrac{1}{2}\ell_2 + \tfrac{1}{4}$$
$$+ \tfrac{1}{2} - \tfrac{3}{4} - \ell_2^2 - \ell_2]\hbar^2$$
$$= (\tfrac{1}{2})\ell_2\hbar^2$$

For the anti-parallel (↑↓) Orbital spins case, equation (A4.1) reduces to

$$(A4.3)\ S_1 \bullet L_2\ (\uparrow\downarrow) = \tfrac{1}{2}[(\ell_2 - \tfrac{1}{2})(\ell_2 - \tfrac{1}{2} + 1) -$$
$$= -\tfrac{3}{4} - \ell_2^2 - \ell_2]\hbar^2$$
$$= -\tfrac{1}{2}[\ell_2 + 1]\hbar^2$$

Note that the quantity $S_1 \bullet L_2$ changes sign in going from the parallel case to the anti-parallel case. In fact their relationship is expressed by equation

$$(A4.4)\ S_1 \bullet L_2\ (\uparrow\downarrow) = -(S_1 \bullet L_2\ (\uparrow\uparrow) + \tfrac{1}{2}\,\hbar^2)$$

Note that there are similar relationships between S_2 and L_1, namely

$$(A5)\ (S_2 + L_1)^2 = (S_2 + L_1)\bullet(S_2 + L_1) =$$
$$= S_2\bullet S_2 + 2S_2\bullet L_1 + L_1\bullet L_1$$
$$= S_2^2 + 2S_2\bullet L_1 + L_1^2$$

Solving for $S_2\bullet L_1$ yields equation

$$(A5.1)\ S_2\bullet L_1 = \tfrac{1}{2}[(S_2 + L_1)^2 - (S_2^2 + L_1^2)]$$

For the parallel (↑↑) orbital spin case, equation (A5.1) reduces to

$$(A5.2)\ S_2\bullet L_1(\uparrow\uparrow) = \tfrac{1}{2}[(\ell_1 + \tfrac{1}{2})(\ell_1 + \tfrac{1}{2} + 1) -$$
$$(\tfrac{1}{2})(\tfrac{1}{2}+1) + \ell_1(\ell_1+1)]\hbar^2$$
$$= \tfrac{1}{2}\,[\ell_1^2 + \tfrac{1}{2}\ell_1 + \ell_1 + \tfrac{1}{2}\ell_1 + \tfrac{1}{4}$$
$$+ \tfrac{1}{2} - \tfrac{3}{4} - \ell_1^2 - \ell_1]\hbar^2$$
$$= (\tfrac{1}{2})\ell_1\hbar^2$$

For the anti-parallel (↑↓) Orbital spins case, equation (A5.1) reduces to

(A5.3) $S_2 \bullet L_1(\uparrow\downarrow) = 1/2[(\ell_1 - \frac{1}{2})(\ell_1 - \frac{1}{2} + 1) -$

$$= -3/4 - \ell_1^2 - \ell_1]\hbar^2$$

$$= -1/2[\ell_1 + 1]\hbar^2$$

Note again, that the quantity $S_2 \bullet L_1$ changes sign in going from the parallel case to the anti-parallel case. In fact their relationship is expressed by equation

(A5.4) $S_2 \bullet L_1 (\uparrow\downarrow) = -(S_2 \bullet L_1 (\uparrow\uparrow) + \hbar^2/2)$

Thus, if the force between identical fermions is dependent on the quantity $S_1 \bullet L_2$, or $S_2 \bullet L_1$ then assume in the case of anti-parallel ($\uparrow\downarrow$) spin and orbital angular momentum, the force is positive (attractive). Thus, in the case of parallel spin and orbital angular momentum, the force is assumed negative (repulsive). These two forces will be referred to as the spin-orbit coupling forces.

Chapter 9

Light Nuclei

Deuterium

It is well known that the deuterium nucleus consists of a neutron and a proton. Moreover, it is known that the deuterium nucleus has an axial spin of $1\hbar$ and is therefore a boson (see glossary for the definition of a boson). Additionally, the deuterium nucleus has no color and has a charge of $+e^+$ which is the same as a proton charge. Experimentally, deuterium is formed when a neutron and proton interact via the nuclear force.

Experimental Explanation of How the Axial Spin of Deuterium Nuclei is 1$\hbar$

The symbolic description of atomic nuclei is written as: $_ZX^A$ where Z is the number of protons, X is the periodic table symbol of the element and A is the number of nucleons (protons + neutrons) in the nucleus.

The following equation describes the formation of deuterium (an isotope of hydrogen, H) observed experimentally

(L1.1) n + p $\rightarrow$ [$_1H^2$] $\rightarrow$ $_1H^2$ + γ

where n is a neutron, p is a proton, $_1H^2$ is deuterium and γ is a photon. Both neutrons and protons have axial spins of ½$\hbar$. Photons have an axial spin of 1$\hbar$. In terms of up (u) and down (d) quarks equation (L1.1) is

(L1.2) d↑u↓d↑ + u↓d↑u↓ $\rightarrow$ [u↑d↓u↑d↓u↑d↓]

70   The Nuclear Force

$$\rightarrow u\uparrow d\downarrow u\uparrow d\uparrow u\downarrow d\uparrow + \gamma\downarrow\downarrow$$

where ↑ means axial spin of $+\frac{1}{2}\hbar$, ↓ means axial spin of $-\frac{1}{2}\hbar$ and ↓↓ means axial spin of $-\hbar$. Here the bracket symbols [ ] mean an excited state or excited configuration. Hence, deuterium has the quark spin configuration of

(L1.3) $_1H^2\uparrow\uparrow = u\uparrow d\downarrow u\uparrow d\uparrow u\downarrow d\uparrow$

because of the conservation of angular momentum (see conservation laws after the glossary) since the photon has a spin of $1\hbar$.

The following figure (Figure 6) gives a pictorial of a deuterium nuclei. It includes the charge, color and spin of each of the three up and down quarks. Recall the up and down quarks attract because of the coulomb force. The three inner down quarks are beginning to fill the first (ground state) down quark energy level. The three outer up quarks are beginning to fill the first (ground state) up quark energy level

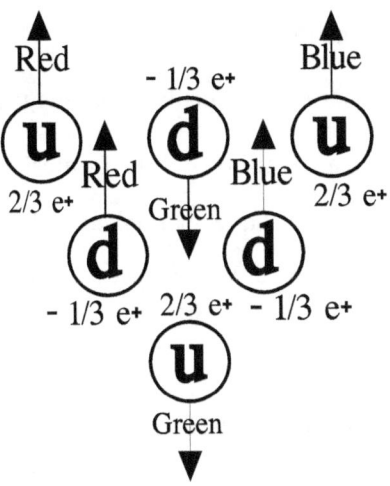

**Figure 6 Deuterium quark
spin, color and charge**

Tritium

The following equation describes the formation of tritium (an isotope of hydrogen, H) observed experimentally

72 The Nuclear Force

(L2.1) $n + {}_1H^2 \rightarrow [\,{}_1H^3\,] \rightarrow {}_1H^3 + \gamma$

where n is a neutron, ${}_1H^2$ is deuterium, ${}_1H^3$ is tritium and γ is a photon. The neutron and the tritium nucleus have axial spins of $½\hbar$. Deuterium (${}_1H^2$) nuclei and Photons have an axial spin of $1\hbar$. In terms of up (u) and down (d) quarks equation (L2.1) is

(L2.2) d↑u↓d↑ + u↑d↓u↑d↑u↓d↑ →
  [d↑u↓d↑u↑d↓u↑d↑u↓d↑]
  → d↓u↑d↓u↑d↓u↑d↑u↓d↑ + γ↓↓

Again the bracket symbols [ ] mean an excited state or excited configuration. Hence, tritium has the quark spin configuration of

(L2.3) ${}_1H^3$↑ = d↓u↑d↓u↑d↓u↑d↑u↓d↑

because of the conservation of angular momentum (see conservation laws after the glossary) since the photon has a spin of $1\hbar$.

Figure 7 gives a pictorial of a tritium nuclei. It includes the charge, color and spin of each of the four up and the five down quarks.

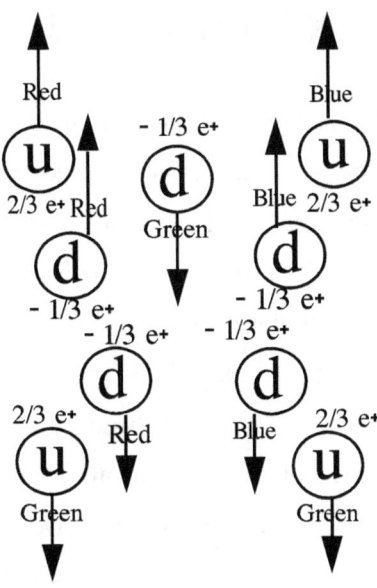

Figure 7 Tritium quark spin, color and charge

Theoretical Explanation of How the Axial Spin of a Deuterium Nucleus is 1$\hbar$

In chapter 8, it was assumed that the spin-spin coupling force for identical colored quarks or fermions composed of colored quarks was attractive for anti-parallel (↑↓) spins and repulsive for parallel (↑↑) spins. Recall that in figure 5, the proton model shows both up quarks have their spins parallel. This does not violate the Pauli exclusion principle (see glossary) since not all their quantum numbers are identical (since their colors are different). This model is also consistent with the assumption that both up quarks repel each other as described by the proton quark dynamics in chapter 4 and the force fields on the proton's up quark of chapter 7. It will now be assumed that if the colored quarks or fermions composed of colored quarks are not identical, then for anti-parallel (↑↓) spins, the spin-spin coupling force is repulsive. Similarly, if the colored quarks or fermions composed of colored quarks are not identical it will be assumed that for parallel spins (↑↑), the spin-spin coupling force is attractive. The following table summarizes all these assumptions about the direction of the color Hooke forces acting between up and down dissimilar colored quarks.

Force Type	Quark 1	Spin 1	Quark 2	Spin 2
Attractive	up	↑	down	↑
Attractive	up	↑	up	↓
Attractive	down	↑	down	↓
Repulsive	up	↑	down	↓
Repulsive	up	↑	up	↑
Repulsive	down	↑	down	↑

Table 1   Spin Dependent Quark Forces

76 The Nuclear Force

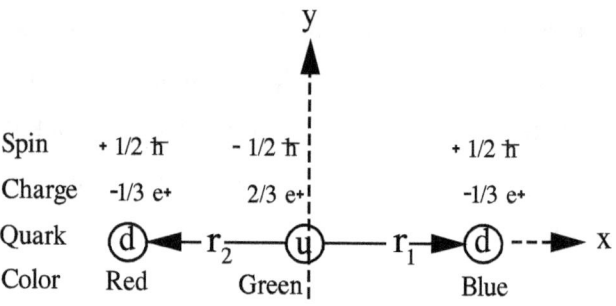

Figure 8   Neutron Coordinate System

This means that quark 1 and quark 2 do not have the same positive color as defined in figure 1 of chapter 1. Recall also that the spin (axial angular momentum) of quarks are either parallel or anti-parallel to each other.

Consider figure 8, above which shows a model for the neutron (similar to figure 5 showing a proton).

Note that both down quarks have parallel spins (↑↑) and since they have different color quantum number, do not violate the Pauli exclusion principle.

Attractive Color Forces

Note that the color force is attractive between two quarks when both quarks are the same species and have anti-parallel (↑↓) spins. Moreover, the color force is attractive between two quarks when the quarks are of different species and have parallel (↑↑) spins.

Repulsive Color Forces

Note that the color force is repulsive between two quarks when both quarks are the same species and have parallel (↑↑) spins. Moreover, the color force is repulsive between two quarks when the quarks are of different species and have antiparallel (↑↓) spins.

Thus, with all the above assumptions on the color forces between quarks in both neutrons and protons yields the picture of figure 6.

Chapter 10

Elementary Particles

In the standard model, particles are categorized as family members and families. Some particles have rest mass (mass while at rest) and some do not. Some particles have charge and some do not. Charge always resides on mass. Recall that charge comes in two (plus or minus) flavors. Particles also have intrinsic spin angular momentum expressed as either half integral values or whole integral values of Planck's rationalized constant, $\hbar$. Matter particles called fermions have half integral spins while force particles called bosons have integral spins.

Particle Generator

If it is assumed that the infinitesimal charge with respect to the infinitesimal mass of any particle is constant, then,

(EP1.1) $d^2Q_{ij}/dM_{ij}^2 = 0$

where Q is the charge, M is the mass, j is the family and i is the family member. Thus,

(EP1.2) $dQ_{ij}/dM_{ij} = A_{ij}$

where A_{ij} are constants of the first integrations. So,

(EP1.3) $Q_{ij} = A_{ij}M_{ij} + B_{ij}$

and where the B_{ij} are also constants of the second integrations.

If it is assumed that the infinitesimal spin angular momentum with respect to the infinitesimal mass of any particle is constant, then,

(EP2.1) $d^2L_{ij}/dM_{ij}^2 = 0$

where L is the spin, M is the mass, j is the family and i is the family member. Thus,

(EP2.2) $dL_{ij}/dM_{ij} = C_{ij}$

where C_{ij} are constants of the first integrations. So,

(EP2.3) $L_{ij} = C_{ij}M_{ij} + D_{ij}$

and where the D_{ij} are also constants of the second integration.

Matter Particles

Matter particles consist of three quark families of four particles each.

Quark Families

The up family consists of the up and down quarks, the electron and electron neutrino. The up quark has a charge of 2/3 that of a positron (2/3 e+). The down quark has a negative charge of 1/3 that of a positron (–1/3 e+). The electron has the opposite charge of a positron (–1e+). The electron neutrino has no charge. The proton consists of two up quarks and one down quark. The neutron consists of two down quarks and one up quark. Recall that all nuclei consist of protons and neutrons.

The strange family consists of the strange and charmed quarks, the muon and the muon neutrino. The strange quark has a charge of 2/3 that of a positron (2/3 e+). The charmed quark has a negative charge of 1/3 that of a positron (–1/3 e+). The muon has the opposite charge of a positron (–1e+). The muon neutrino has no charge.

The top family consists of the top and bottom quarks, the tauon , and the tauon neutrino. The top quark has a charge of 2/3 that of a positron (2/3 e+). The bottom quark has a negative charge of 1/3 that of a positron (–1/3 e+). The tauon has the opposite charge of a positron (–1e+). The tauon neutrino has no charge.

The Up Family

Let $j = 1$ stand for the up family while $i = 1, 2, 3$ stand for the up quark, down quark and electron. $i = 4$ stands for the electron neutrino. Equation (EP1.3) becomes

(EP1.4) $Q_{i1} = A_{i1}M_{i1} + B_{i1}$

Here the $B_{i1} = 0$ for $i = 1, 2, 3$ and Q_{i1} is the up quark charge, down quark charge and electron charge as $i = 1, 2, 3$. Thus equation (EP1.4) becomes

(EP1.5) $Q_{i1} = A_{i1}M_{i1}$ ($i = 1, 2, 3$)

The electron neutrino has no charge so that $Q_{41} = 0$ and if $B_{41} \neq 0$ the mass of the electron neutrino by equation (EP1.4) reduces to

(EP1.6) $M_{41} = -B_{41}/A_{41}$

Note that if $B_{41} = 0$, then the electron neutrino has zero rest mass.

The spin angular momentum of the four members of the up family is known to be $\hbar/2$ where $\hbar$ is Planck's rationalized constant. Let $j = 1$ stand for the up family while $i = 1, 2, 3, 4$ stand for the up quark, down quark, electron and electron neutrino. Equation (EP2.3) becomes

(EP2.4) $L_{i1} = C_{i1}M_{i1} + D_{i1}$ ($i = 1, 2, 3, 4$)

if the electron neutrino has zero rest mass, then $M_{41} = 0$ and $L_{41} = D_{41} = \hbar/2$. If the electron neutrino has non zero mass then $D_{41} = 0$ and the spin of the electron neutrino is $L_{41} = C_{41}M_{41} = \hbar/2$.

The Strange Family

Let $j = 2$ stand for the strange family while $i = 1, 2, 3$ stand for the strange quark, charmed quark and muon. $i = 4$ stands for the muon neutrino. Equation (EP1.3) becomes

(EP1.7) $Q_{i2} = A_{i2}M_{i2} + B_{i2}$

Here the $B_{i2} = 0$ for $i = 1, 2, 3$ and Q_{i2} is the strange quark charge, charmed quark charge and muon charge as $i = 1, 2, 3$. Thus equation (EP1.7) becomes

(EP1.8) $Q_{i2} = A_{i2}M_{i2}$ ($i = 1, 2, 3$)

The electron neutrino has no charge so that $Q_{42} = 0$ and if $B_{42} \neq 0$ the mass of the muon neutrino by equation (EP1.7) reduces to

(EP1.9) $M_{42} = -B_{42}/A_{42}$

Note that if $B_{42} = 0$, then the muon neutrino has zero mass.

The spin angular momentum of the four members of the strange family is known to be $\hbar/2$ where $\hbar$ is Planck's rationalized constant. Let j = 2 stand for the strange family while i = 1, 2, 3, 4 stand for the strange quark, charmed quark, muon and muon neutrino. Equation (EP2.3) becomes

(EP2.5) $L_{i2} = C_{i2}M_{i2} + D_{i2}$ (i = 1, 2, 3, 4)

if the muon neutrino has zero rest mass, then $M_{42} = 0$ and $L_{42} = D_{42} = \hbar/2$. If the muon neutrino has non zero mass then $D_{42} = 0$ and the spin of the electron neutrino is $L_{42} = C_{42}M_{42} = \hbar/2$.

Elementary Particles 87

The Top Family

Let $j = 3$ stand for the top family while $i = 1, 2, 3$ stand for the up quark, bottom quark and tauon. $i = 4$ stands for the tauon neutrino. Equation (EP1.3) becomes

(EP1.10) $Q_{i3} = A_{i3}M_{i3} + B_{i3}$

Here the $B_{i3} = 0$ for $i = 1, 2, 3$ and Q_{i3} is the top quark charge, bottom quark charge and tauon charge as $i = 1, 2, 3$. Thus equation (EP1.10) becomes

(EP1.11) $Q_{i3} = A_{i3}M_{i3}$ ($i = 1, 2, 3$)

The tauon neutrino has no charge so that $Q_{43} = 0$ and if $B_{43} \neq 0$ the mass of the tauon neutrino by equation (EP1.10) reduces to

(EP1.12) $M_{43} = -B_{43}/A_{43}$

Note that if $B_{43} = 0$, then the tauon neutrino has zero mass.

The spin angular momentum of the four members of the top family is known to be $\hbar/2$ where $\hbar$ is Planck's rationalized constant. Let $j = 3$ stand for the strange family while $i = 1, 2, 3, 4$ stand for the top quark, bottom quark, tauon and tauon neutrino. Equation (EP2.3) becomes

(EP2.6) $L_{i3} = C_{i3}M_{i3} + D_{i3}$ ($i = 1, 2, 3, 4$)

if the tauon neutrino has zero rest mass, then $M_{43} = 0$ and $L_{43} = D_{43} = \hbar/2$. If the tauon neutrino has non zero mass then $D_{43} = 0$ and the spin of the tauon neutrino is $L_{43} = C_{43}M_{43} = \hbar/2$.

Force Particles

Force particles consist of four families which account for all the four forces of nature.

These are the strong force carried by the gluon family. There are eight gluons. Gluons are responsible for the forces between quarks.

The electromagnetic force is carried by the photon family. There are two photons. Positive mass photons cause electromagnetic forces associatied with charge residing on matter particles. It is postulated that negative mass photons cause electromagnetic forces associated with charge residing on negative anti-matter particles.

The weak force is carried by the weakon family. There are three weakon particles. The omega minus member is responsible for beta (electron) radioactive decay. The omega plus member is responsible for alpha (helium nuclei) radioactive decay. The neutral zeta member is responsible for gamma (photons) radioactive decay.

The gravitational force is carried by the graviton family. It is postulated that there are two gravitons. In a book entitled "The Nature of the First Cause" a theory put forth by the author showed that positive gravitons cause a repulsive force between positive mass and negative anti-mass.

Negative anti-gravitons cause the attractive forces between positive masses and the attractive forces between negative anti-masses.

The Gluon Family

Let $j = 4$ stand for the gluon family. i will range from 1 to 8 for the eight gluons. Since the gluons do not carry charge, equation

(EP1.3) $Q_{ij} = A_{ij}M_{ij} + B_{ij}$

reduces to

(EP1.13) $M_{i4} = -B_{i4}/A_{i4}$ ($i = 1, 8$)

where

(EP1.13.1) $Q_{i4} = 0$ ($i = 1, 8$)

The spin angular momentum of each gluon is assumed to be $\hbar$ by the standard model. Equation (EP2.3) becomes

(EP2.7) $L_{i4} = C_{i4}M_{i4} + D_{i4}$ ($i = 1, 8$)

Here the $D_{i4} = 0$ ($i = 1, 8$) with

(EP2.7.1) $L_{i4} = C_{i4}M_{i4} = \hbar$ (i = 1, 8)

The Photon Family

Let j = 5 stand for the photon family. i will range from 1 to 2 for the two photons. Since photons do not carry charge, equation

(EP1.3) $Q_{ij} = A_{ij}M_{ij} + B_{ij}$

reduces to

(EP1.14) $M_{i5} = -B_{i5}/A_{i5}$ (i = 1, 2)

where

(EP1.14.1) $Q_{i5} = 0$ (i = 1, 2)

The spin angular momentum of each photon is assumed to be $\hbar$ by the standard model. Equation (EP2.3) becomes

92　The Nuclear Force

(EP2.8) $L_{i5} = C_{i5}M_{i5} + D_{i5}$ (i = 1, 2)

Here the $D_{i5} = 0$ (i = 1, 2) with

(EP2.8.1) $L_{i5} = C_{i5}M_{i5} = \hbar$ (i = 1, 2)

The Weakon Family

Let j = 6 stand for the weakon family. i will range from 1 to 3 for the three weakons. Since the omega plus and the omega minus carry positive and negative charge, equation

(EP1.3) $Q_{ij} = A_{ij}M_{ij} + B_{ij}$

reduces to

(EP1.15) $Q_{i6} = A_{i6}M_{i6}$ (i = 1,2)

where

(EP1.15.1) $B_{ij} = 0$ $(i = 1,2)$

for the omega plus and omega minus. The zeta zero carries no charge and equation

(EP1.3) $Q_{ij} = A_{ij}M_{ij} + B_{ij}$

reduces to

(EP1.16) $M_{36} = -B_{36}/A_{36}$

for the zeta zero where

(EP1.16.1) $Q_{36} = 0$

The spin angular momentum of each weakon is assumed to be $\hbar$ by the standard model. Equation (EP2.3) becomes

(EP2.9) $L_{i6} = C_{i6}M_{i6} + D_{i6}$ $(i = 1, 3)$

Here the $D_{i6} = 0$ ($i = 1, 3$) with

(EP2.9.1) $L_{i6} = C_{i6}M_{i6} = \hbar$ ($i = 1, 3$)

The Graviton Family

Let $j = 7$ stand for the graviton family. i will range from 1 to 2 for the two gravitons. Since the gravitons do not carry charge, equation

(EP1.3) $Q_{ij} = A_{ij}M_{ij} + B_{ij}$

reduces to

(EP1.17) $M_{i7} = -B_{i7}/A_{i7}$ ($i = 1, 2$)

where

(EP1.17.1) $Q_{i7} = 0$ ($i = 1, 2$)

The spin angular momentum of each graviton is assumed to be $2\hbar$ by the standard model. Equation (EP2.3) becomes

(EP2.10) $L_{i7} = C_{i7}M_{i7} + D_{i7}$ ($i = 1, 2$)

Here the $D_{i7} = 0$ ($i = 1, 2$) with

(EP2.10.1) $L_{i6} = C_{i6}M_{i6} = 2\hbar$ ($i = 1, 2$)

It is interesting to note that a single equation (EP3.1 below) can generate all the known elementary particles and their properties. Equations (EP1.1) and (EP2.1) may be combined (since $0*0 = 0$) into

(EP3.1) $(d^2Q_{ij}/dM_{ij}^2)(d^2L_{ij}/dM_{ij}^2) \equiv 0$

Chapter 11

Cold Fusion

There is at least two cold fusion prototype reactors that have emerged and have been recently reported in the literature (Infinite Energy volume 21, issue 122). Recall that in 1989, Stanley Pons and Martin Fleischman announced that they had discovered "cold fusion" in an electrolytic cell with an electrode made of palladium in a solution of heavy water (deuterium oxide).

Many USA universities including MIT initially tried to replicate the experiment without success. The american physical society claimed that the experiment was flawed and that Pons and Fleischmann had reported inaccurate results. The USA patent office refused to review any inventions which contained the words "cold fusion".

Subsequent experiments showed that the experiment had been valid. MIT is now offering classes on LENR (low energy nuclear reactions) which has replaced the term "cold fusion". Experimentally, two cold fusion reactions have been successfully investigated. One involves nickel and hydrogen (one proton in the nucleus). The other involves palladium and deuterium (proton and neutron in nucleus).

Preliminary Considerations

It is known chemically that palladium can absorb deuterium and nickel can absorb hydrogen. Both palladium and nickel exist naturally as several stable isotopes.

The following is the standard notation for representing any element. The notation is $_ZX^A$ where Z is the number of protons in the nucleus. X is the symbol for the element and A is the atomic number (protons plus neutrons).

Nickel (Ni) has five stable isotopes. These are $_{28}Ni^{58}$, $_{28}Ni^{60}$, $_{28}Ni^{61}$, $_{28}Ni^{62}$ and $_{28}Ni^{64}$. The approximate abundant percentages occuring in natural nickel are respectively, 68%, 26%, 1%, 4% and 1%.

Palladium (Pd) has six stable isotopes. These are $_{46}Pd^{102}$, $_{46}Pd^{104}$, $_{46}Pd^{105}$, $_{46}Pd^{106}$, $_{46}Pd^{108}$ and $_{46}Pd^{110}$. The approximate abundant percentages occuring in natural palladium are respectively, 1%, 11%, 22%, 27%, 27% and 12%.

The following preliminary equations are as follows:

(FU1.0) $_1H^1 + e^- \rightarrow n + \nu$

where $_1H^1$ is a proton, e^- is an electron, n is a neutron and ν is a neutrino.

(FU1.1) $_1H^1 + n \rightarrow {_1H^2} + \gamma$

where $_1H^2$ is deuterium and γ is a photon.

(FU1.2) $n + {_1H^2} \rightarrow [_1H^3] \rightarrow {_1H^3} + \gamma$

where $_1H^3$ is tritium. The closed brackets [ ] mean an excited state or compound nucleus of whatever is inside the brackets.

Nickel and Hydrogen

Andre Rossi was the first to develop a prototype reactor called the E-Cat (energy catalyzer). It basically consisted of a nickel cathode immersed in a hydrogen pressurized vessel. A "loading phase" in which the nickel absorbed hydrogen was observed. Next, a current was run through the nickel (with nickel as the anode). The nickel heated up beyond anything that could have been a chemical reaction. Both heat (photons) and copper was observed to form during this "transmutation" phase. Rossi has reported that after the "heating", roughly 1/3 of the nickel had been converted to copper. The following set of nuclear reactions involving neutrons (which are produced via equation (FU1.0 $_1H^1 + e^- \rightarrow n + \nu$) above are proposed to occur.

(FU2.0) $_{28}Ni^{60} + n \rightarrow [_{28}Ni^{61}] \rightarrow {}_{28}Ni^{61} + \gamma$

(FU2.1) $_{28}Ni^{61} + n \rightarrow [_{28}Ni^{62}] \rightarrow {}_{28}Ni^{62} + \gamma$

Note that heat (photons) have been produced as nickel has gained in atomic number.

(FU2.2) $_{28}Ni^{62} + {}_1H^1 \rightarrow [_{29}Cu^{63}] \rightarrow {}_{29}Cu^{63} + \gamma$

where Cu is copper. Also copper is produced by the following reaction

(FU2.3) $_{28}Ni^{64} + {}_1H^1 \rightarrow [_{29}Cu^{65}] \rightarrow {}_{29}Cu^{65} + \gamma$

Both stable isotopes of copper are produced, namely $_{29}Cu^{63}$ and $_{29}Cu^{65}$, as well as heat. Note that

$_{28}Ni^{60}$ represents 26% of naturally occurring nickel. $_{28}Ni^{62}$ and $_{28}Ni^{64}$ represent another 5% of naturally occurring nickel. Together, 31% (26% + 5%) of the nickel has been converted to copper. This agrees with Rossi's statement that roughly 1/3 of the nickel in his E-Cat reactor had been converted to copper!

Recently, Andrea Rossi modified his reactor and instead of pressurizing his vessel with hydrogen, he added the compound LiAlH4 which when heated, chemically produced hydrogen, aluminum (Al) and lithium (Li). Lithium occurs as two isotopes namely, $_3Li^6$ and $_3Li^7$. Percentage abundances of $_3Li^7$ occur at 92% and $_3Li^6$ at 8%. Proposed nuclear reactions are as follows:

(FU3.0) $_3Li^6 + n \rightarrow [_3Li^7] \rightarrow {}_1H^3 + {}_2He^4$

(FU3.1) $_3Li^6 + {}_1H^2 \rightarrow [_4Be^8] \rightarrow {}_2He^4 + {}_2He^4$

where Be is beryllium and He is helium. Deuterium ($_1H^2$) can be produced by equations (FU1.1) $_1H^1 + n \rightarrow {}_1H^2 + \gamma$) above. Rossi has stated that his reactor has produced both tritium ($1H^3$) as well as helium.

Palladium and Deuterium

Pons and Fleischmann used palladium as the cathode in their electrolytic cell containing heavy water (D_2O). After the palladium had chemically absorbed and was saturated with deuterium, the poles of the cell were reversed effectively making the palladium the anode. Within the palladium lattice, deuterium nuclei were driven to interact via a quantum mechanical effect called tunneling. The following initial nuclear reactions include:

(FU4.0) $_1H^2 + {}_1H^2 \rightarrow [{}_2He^4] \rightarrow {}_1H^3 + {}_1H^1$

(FU4.1) $_1H^2 + {}_1H^2 \rightarrow [{}_2He^4] \rightarrow {}_2He^3 + n$

With the production of neutrons via equation (FU4.1) the following reactions may occur.

(FU4.2) $_{46}Pd^{104} + n \rightarrow [{}_{46}Pd^{105}] \rightarrow {}_{46}Pd^{105} + \gamma$

(FU4.3) $_{46}Pd^{105} + n \rightarrow [{}_{46}Pd^{106}] \rightarrow {}_{46}Pd^{106} + \gamma$

Note that heat is produced as the palladium gains in atomic number. With the production of protons via equation (FU4.0 $_1H^2 + {_1}H^2 \rightarrow [_2He^4] \rightarrow {_1}H^3 + {_1}H^1$) the following reactions may occur.

(FU4.4) $_{46}Pd^{106} + {_1}H^1 \rightarrow [_{47}Ag^{107}] \rightarrow {_{47}}Ag^{107} + \gamma$

(FU4.5) $_{46}Pd^{108} + {_1}H^1 \rightarrow [_{47}Ag^{109}] \rightarrow {_{47}}Ag^{109} + \gamma$

Note that heat and silver (Ag) are produced. Palladium can also interact with deuterium in the following reactions:

(FU4.6) $_{46}Pd^{105} + {_1}H^2 \rightarrow [_{47}Ag^{107}] \rightarrow {_{45}}Rh^{103} + {_2}He^4$

Note that in this reaction, both helium and rhodium are produced.

(FU4.7) $_{46}Pd^{106} + {_1}H^2 \rightarrow [_{47}Ag^{108}] \rightarrow {_{47}}Ag^{107} + n$

(FU4.8) $_{46}Pd^{108} + {_1}H^2 \rightarrow [_{47}Ag^{110}] \rightarrow {_{47}}Ag^{109} + n$

(FU4.9) $_{46}Pd^{104} + {}_1H^2 \rightarrow [{}_{47}Ag^{106}] \rightarrow {}_{46}Pd^{105} + {}_1H^1$

(FU4.10) $_{46}Pd^{105} + {}_1H^2 \rightarrow [{}_{47}Ag^{107}] \rightarrow {}_{46}Pd^{106} + {}_1H^1$

Note that in these last four reactions, silver (Ag), protons and neutrons are produced.

All the nuclear reactions listed above for both the nickel hydrogen and the palladium deuterium reactor prototypes attempt to explain the excess energy that has been reported. It must be mentioned that this set of reactions does not represent all the possibilities. Surely, cold fusion reactors will not be limited to the two which have been presented in this chapter.

GLOSSARY

Bose-Einstein Statistics: Mechanics describing the statistical behavior of a group of Bosons.

Boson: A particle which has its spin angular momentum equal to an integral number of Planck's constant divided by 2π, $\hbar$.

Coulomb's Law: The force between two charges, q_1 and q_2 is proportional to the product of the charges and inversely proportional to the square of the distance, r between their charge centers. The constant of proportionality, K_C is defined to be Coulomb's constant and its value is 8.99 X 10^9 nt-m^2/coul^2. The force is always in a direction along the line passing through both charge centers. If the charges have opposite signs, the force is attractive (positive). If the charges have the same signs, the force is repulsive (negative). $|F| = K_C q_1 q_2 / r^2$

Del Operator: An operator, ∇ equal to the vector whose xyz components are $\nabla = (\partial/\partial x, \partial/\partial y, \partial/\partial z)$.

Eigenstate: A wave function Ψ (the eigenstate), such that if an operator, $\hat{O}$ multiplied by the wave function, Ψ yields a number, E (the eigenvalue) multiplied by the wave function. If $\hat{O}\Psi = E\Psi$, then Ψ is the eigenstate.

Eigenvalue: If an operator, $\hat{O}$ multiplied by the wave function, Ψ yields a number, E (the eigenvalue) multiplied by the wave function. If $\hat{O}\Psi = E\Psi$, then E is the eigenvalue.

Electron Volt: An electron volt (ev)(1.602×10^{-19} joules) is the kinetic energy an electron gains by being propelled a distance of one meter by an electrical field of strength, one volt per meter.

Fermi-Dirac Statistics: Mechanics describing the statistical behavior of a group of Fermions.

Fermion: A particle which has its spin angular momentum equal to a half integral number of Planck's constant divided by 2π, $\hbar/2$.

Feynman Diagram: A standard tool used for the depiction of a quantum mechanical process. Wavy arrows stand for zero rest mass bosons including photons. Solid arrows stand for non-zero rest mass Fermions. Arrows are in the same direction as the flow of time.

Graviton: The elementary boson with a spin of 2 associated with the gravitational field. It is postulated that there are two gravitons having positive and negative energy. Negative gravitons operate between attractive positive masses., while positive gravitons operate between repelling positive mass and negative mass.

Hamiltonian Operator: An operator, $\hat{H}$ which when multiplied by the wave function, Ψ produces the total energy eigenvalue, E. $\hat{H}\Psi = E\Psi$

Helicity: The component of a particles spin angular momentum in the direction of the particle's velocity vector.

Hooke's Law: The force, F on a particle suspended by a spring is proportional to the negative displacement, x of the spring. The constant of proportionality is the spring constant k. $F = -kx$

Laplacian Operator: An operator which is the dot product of the Del = $\nabla = (\partial/\partial x, \partial/\partial y, \partial/\partial z)$ with itself or $\nabla \cdot \nabla = \nabla^2 = (\partial^2/\partial x^2, \partial^2/\partial y^2, \partial^2/\partial z^2)$.

Lepton: A Fermion particle associated with the three families of matter particles. Each family contains two leptons, one of which has a negative electronic charge and the other which has no charge. See also standard model.

MEV: a million electron volts

Neutrino: A chargeless Fermion particle (lepton) associated with beta (electron) radioactivity. See also standard model.

Newton's Universal Gravitational Law: The force between any two masses, m_1 and m_2 is proportional to the product of their masses and inversely proportional to the square of the distance, r between their centers. The constant of proportionality is Newton's Universal Gravitational Constant, $G = 6.67 \times 10^{-11}$ newton-meters^2/kilograms^2. $|F| = Gm_1m_2/r^2$

Pauli Exclusion Principle: No two Fermions can have all their quantum numbers the same.

Photon: The Boson particle associated with electromagnetic radiation and forces between charges.

Planck Acceleration: $A_P = c/T_P = 5.56 \times 10^{35}$ meters/second^2

Planck Boson Spin: $\hbar = 1.0552 \times 10^{-34}$ joule–seconds

Planck's Constant: $h = 6.63 \times 10^{-34}$ joule-seconds

Planck's Rationalized Constant: $h/2\pi = \hbar = 1.0552 \times 10^{-34}$ joule-seconds

Planck Energy: $E_P = M_P c^2 = 1.96 \times 10^9$ joules

Planck Fermion Spin: $\hbar/2 = 5.276 \times 10^{-35}$ joule–seconds

Planck Force: $F_P = GM_P^2/L_P^2 = M_P c/T_P = c^4/G = 1.21 \times 10^{44}$ newtons

Planck Frequency: $W_P = M_P c^2/\hbar = 1.857 \times 10^{44}$ hertz

Planck Length: $L_P = (\hbar G/c^3)^{1/2} = 1.616 \times 10^{-35}$ meters

Planck Mass: $M_P = (\hbar c/G)^{1/2} = 2.177 \times 10^{-8}$ kilograms

Planck Momentum: $P_P = M_P c = 6.53$ kilogram–meters/second

Planck Time: $T_P = (\hbar G/c^5)^{1/2} = 5.391 \times 10^{-44}$ seconds

Speed of Light in Vacuum: $c = 3.00 \times 10^8$ meters/second

STANDARD MODEL: All matter is composed of three families of four fermion particles each. The first two members are quarks and the last two members are leptons. These matter families are:

Up family (up quark, down quark, electron, electron anti–neutrino)

Charmed family (charmed quark, strange quark, muon, muon anti–neutrino)

Top family (top quark, bottom quark, tauon, tauon anti–neutrino)

All field energy is composed of three families of three boson particles each.

The GPG family includes the gluons responsible for the force that holds the quarks in both neutrons (2 down quarks & 1 up quark) and protons (2 up quarks & 1 down quark) together. The photon is responsible for forces between charges. The graviton is responsible for forces between masses.

The pion family includes the 3 pions responsible for holding neutrons and protons of atomic nuclei together.

The weakon family includes the 3 weakons responsible for forces causing radioactive decay. In summary, these three field families are:

GPG family (gluon, photon, graviton), pion family (pi plus, pi minus, pi zero) and the weakon family (omega plus, omega minus, zeta zero).

All particles also have a partner which is called its anti–particle. Anti–particles have the same mass as the particles but the sign of their charge and helicity (if any) is reversed (See basic elementary particles section).

Wave Function: A mathematical function of space and time, which describes the wave behavior of an energy system.

Fundamental Physical Laws

Preliminary Definitions

I. **Bold** mathematical single letters refer to vectors.

II. The symbol, $i = (-1)^{1/2}$ always occurs as the fourth (time component) of all Einsteinian four dimensional vectors.

III. The symbol, c is the speed of light in vacuum.

1. **Position of an energy system:** Referring to Figure 2, a normal Cartesian coordinate system shows the x,y,z position of the system S at time t.

Newtonian: $\mathbf{r}_N = (x,y,z)$

Einsteinian: $\mathbf{r}_E = (x,y,z,ict)$

2. **Velocity of an energy system:** At time t_2, the position of the system was at position 2 (r_2).

Initially at time t_1, the position of the system was at position 1 (r_1). The average velocity of the system is the distance traversed by the system in moving from position 1 to position 2 ($r_2 - r_1$) divided by the time it took for the system to move between the two positions ($t_2 - t_1$). The direction of the velocity is from position 1 to position 2. The instantaneous velocity **v** is realized by letting t_2 approach t_1. Mathematically, the instantaneous velocity of a system is a vector quantity.

$\mathbf{v} = \lim$ as $t_2 \to t_1$ of $[(\mathbf{r_2} - \mathbf{r_1})/(t_2 - t_1)]$ or

$\mathbf{v} = d\mathbf{r}/dt$

Newtonian: $\mathbf{v}_N = (v_x, v_y, v_z)$

Einsteinian: $\mathbf{v}_E = (v_x, v_y, v_z, ic)$

3. **Acceleration of an energy system:** At time t_2, the velocity of the system was (v_2). Initially at time t_1, the velocity of the system was (v_1). The average acceleration of the system is the change in the

velocity of the system in going from v_1 to v_2, $(v_2 - v_1)$ divided by the time it took for the system to go from v_1 to v_2, $(t_2 - t_1)$. The direction of the acceleration is from v_1 to v_2. The instantaneous acceleration **a** is realized by letting t_2 approach t_1. Mathematically, the acceleration of a system is a vector quantity.

$$\mathbf{a} = \lim_{t_2 \to t_1} [(\mathbf{v_2} - \mathbf{v_1})/(t_2 - t_1)] \quad \text{or}$$

$$\mathbf{a} = d\mathbf{v}/dt$$

Newtonian: $\mathbf{a}_N = (a_x, a_y, a_z)$

Einsteinian: $\mathbf{a}_E = (a_x, a_y, a_z, 0)$

4. **Momentum of an energy system:** The product of the system's mass and its velocity **v** is called its momentum and denoted by **p**. It is a vector quantity having direction **v**.

$$\mathbf{p} = m\mathbf{v}$$

Newtonian: $\mathbf{p}_N = (p_x, p_y, p_z)$

Einsteinian: $\mathbf{p}_E = (p_x, p_y, p_z, iE/c)$

where E is the total energy = mc^2.

5. **Force on an energy system:** The instantaneous rate of change of a system's momentum with respect to time. Its definition is similar to the definition of velocity. At time t_2 it has momentum 2. At time t_1, it had momentum 1. The average force is the difference in momentum (momentum 2 − momentum 1) divided by the time difference ($t_2 - t_1$). The instantaneous rate is realized when t_2 approaches t_1.

$\mathbf{F}$ = lim as $t_2 \rightarrow t_1$ of $[((m\mathbf{v})_2 - (m\mathbf{v})_1)/(t_2 - t_1)]$ or

$\mathbf{F} = d(m\mathbf{v})/dt$

Newtonian: $\mathbf{F}_N = m_0 d\mathbf{v}/dt = m_0 \mathbf{a}$
where m_0 is the rest mass and $\mathbf{a}$ is the acceleration.

Einsteinian: $\mathbf{F}_E = d(m\mathbf{v})/dt = md\mathbf{v}/dt + \mathbf{v}dm/dt$
$= m_0\mathbf{a}(1-(v/c)^2)^{-3/2} = (c^2/v)dm/dt$

where **v** is m's velocity and **a** is m's acceleration.

6. **Mass density:** Mass m, per unit volume V.

Average mass density = ρ_{mavg} = m/V

Instantaneous mass density = ρ_m = dm/dV

7. **Pressure on a surface**: The applied force F, per unit surface area, A.

Average pressure = P_{avg} = F/A

Instantaneous pressure = P = dF/dA

8. **Angular momentum of an energy system:** Let the vector from the origin to the position of the system be called the position vector (**r**). The angular momentum of the system (**L**) is then the

ordinary vector cross product (**x**), of the position vector with the system's momentum vector (**mv**).

L = r x mv

9. **Charge density:** Amount of charge q, per unit volume V.

Average charge density = ρ_{qavg} = q/V

Instantaneous charge density = ρ_q = dq/dV

10. **Electrical current:** The instantaneous change of charge q with respect to time.

i = lim as $t_2 \rightarrow t_1$ of $[(q_2 - q_1)/(t_2 - t_1)]$ or

i = dq/dt

11. **Electrical current density:** The electrical current i per unit cross sectional area A of conductor. The unit vector **u** has a direction of the

current i along the conductor perpendicular to the cross sectional area.

$\mathbf{J}_i = \mathbf{u}i/A$ where $i = dq/dt$

In a conductor with conductivity σ_c, the current is in the direction of the electric field **E** and the electrical current density is the product of the conductivity and electric field.

$\mathbf{J}_i = \sigma_c \mathbf{E}$

12. **Mole**: The mass of Avogadro's number of identical molecules or Avogadro's number of identical atoms expressed in grams. One mole of molecules is the molecular weight of the molecule expressed in grams. One mole of atoms is the atomic weight of the atom expressed in grams.

Mechanical Laws

1. **Newton's Laws of Motion:**

1.1 A body will remain at rest or in motion at a constant velocity unless acted on by an unbalanced external force.

1.2 The force on a body is proportional to its acceleration and the constant of proportionality is the rest mass (when the body is at rest), m_0 of the body.

$\mathbf{F} = m_0 \mathbf{a}$

(Newton was unaware that mass is a function of its velocity.)

1.3 The force of one body on a second body is equal and opposite to the force of the second body on the first body or for every action, there is an equal and opposite reaction.

$\mathbf{F}_{12} = -\mathbf{F}_{21}$

1.4 Newton's Universal Law of Gravitation says that any two energy systems having mass attract each other with a force (**F**) proportional to the product of their masses m_1 and m_2 and inversely proportional to the square of the distance (r) between their mass centers. The force is in a direction between the centers of m_1 and m_2, causing them to attract one another and is denoted by the unit vector $\mathbf{r}_u$. G is the constant of proportionality known as Newton's Gravitational Constant. This force is

$\mathbf{F} = \mathbf{r}_u G m_1 m_2 / r^2$

deriving the gravitational potential energy, V between m_1 and m_2 as

$V = -G m_1 m_2 / r$

Newtonian: Mass is the cause of the gravitational field.

Einsteinian: Mass energy and momentum warp four dimensional spacetime into a gravitational field.

2. **Quantum Mechanical Laws**:

2.1 An energy system may be described by a wave function. The total energy operator $\hat{H}$ (known as the Hamiltonian) operating on the wave function (Ψ) yields the total energy eigenvalue (E) of the system represented by the wave function. Energy eigenvalues (E) are the allowable energy states that the system may assume. Similarly, other operators operating on the wave function yield other information (such as the spin, momentum, angular momentum, etc.) about the system.

$\hat{H}\Psi = E\Psi$

2.2 The square of the wave function $\Psi^*\Psi$, multiplied by a infinitesimal volume d^3r is equal to the infinitesimal probability dP, that a system specified by Ψ, is located within that volume.

$$dP = \Psi^*\Psi d^3r$$

2.3   The probability that an energy system represented by the wave function Ψ, is somewhere in all space is unity, which is the basis for a normalized wave function.

$$P = \int dP = \int \Psi^*\Psi d^3r = 1$$

3. The Heisenberg Uncertainty Principle:

3.1   In an ideal experiment, the product of the standard deviation in the measurement of a system's momentum, Δp and the standard deviation in the measurement of its position, Δr must be greater than a non-zero constant. This constant is Planck's constant divided by four pi ($h/(4\pi) = \hbar/2$) where $\hbar$ is Planck's constant divided by 2π.

$$\Delta p \Delta r \geq \hbar/2$$

This means that an energy system's position and momentum cannot be known simultaneously.

3.2 Another expression of the Heisenberg uncertainty principle is:

$\Delta E \Delta t \geq \hbar/2$

where ΔE is the standard deviation in the measurement of a system's energy and Δt is the standard deviation of the measured times that it had that energy. This means that a system's energy and when it had that energy cannot be known simultaneously.

4. **The energy of a photon (E_γ):** an electromagnetic wave's energy is either the product of its frequency ω, and Planck's constant $\hbar$, or its mass m_γ, and the speed of light, c squared.

$E_\gamma = \hbar\omega = m_\gamma c^2$

5. **De Broglie's relationship:** which expresses that the wavelength of a particle λ is inversely proportional to its momentum. The constant of proportionality is Planck's constant, h.

$$\lambda = h/mv$$

and is sometimes written as

$$\lambdabar = \hbar/mv$$

where $\lambdabar = \lambda/(2\pi)$ and $\hbar = h/(2\pi)$

6. **Einstein Laws of Special Relativity:** The first four relativistic laws are derived by assuming that the velocity of light c, is independent of the velocity of the source of light as well as the velocity of the observer.

6.1 A system's mass m increases if it is moving with a velocity v compared to the velocity of light c, in vacuum. Initially when the system had a velocity of zero, its rest mass is m_0.

$$m = m_0 \left(1 - (v/c)^2\right)^{-1/2}$$

6.2 A system's length ℓ decreases if it is moving with a velocity v compared to the velocity of light c, in vacuum. Initially when the system had a velocity of zero, its rest length is ℓ_0. ℓ is in the direction of the velocity.

$$\ell = \ell_0 (1 - (v/c)^2)^{1/2}$$

6.3 A system's clock time length, t slows (stretches) if it is moving with a velocity v compared to the velocity of light c, in vacuum. Initially when the system was at rest (had a velocity of zero), it had a clock time length of t_0.

$$t = t_0 \left(1 - (v/c)^2\right)^{-1/2}$$

6.4 The total mechanical energy E of a system containing mass is the product of its mass m and the square of the velocity of light c.

$$E = mc^2$$

where $m = m_0(1 - (v/c)^2)^{-1/2}$ is dependent on its velocity v. m_0 (rest mass) is its mass when $v = 0$.

6.5 The relativistic kinetic energy T, of a system in motion is the difference (between its mass in motion less its rest mass) times the velocity of light c, squared.

Einsteinian: $T_E = (m - m_0)c^2$

where $m = m_0(1 - (v/c)^2)^{-1/2}$ is dependent on its velocity, v. For small velocities compared to the velocity of light, the Einsteinian kinetic energy reduces to the Newtonian kinetic energy as a first order approximation. For $v \ll c$, $(m - m_0)c^2 \cong (½)m_0v^2$

Newtonian: $T_N = (½)m_0v^2$

7. Laws of Thermodynamics

7.1 The first law of thermodynamics says that within a closed (isolated) system an amount of heat added to the system dQ results in an increase in its internal energy dU and an amount of work done, dW. Usually, dU results in an increase in internal temperature while dW results in a change in volume dV against a constant pressure p. This also means that energy is conserved for a closed system.

$dQ = dU + dW$ where $dW = pdV$

7.2 The second law of thermodynamics says that a change in the entropy dS of a system undergoing a reversible process is defined to be the amount of heat added dQ divided by its temperature T. If the process is irreversible, then the entropy is always greater than the amount of heat added divided by its temperature.

$dS \geq dQ/T$

where the equality implies reversibility and the greater than symbol (>) implies irreversibility.

7.3 The perfect gas law says that the gas pressure p multiplied by the volume of gas V is proportional to the number of moles n of gas multiplied by the absolute temperature T of the gas. The constant of proportionality R is known as the universal gas constant.

$pV = nRT$

7.4 The fundamental law of heat conduction says that the rate of heat flow dQ/dt across a infinitely thin slab dx of material perpendicular to the surface of the slab is proportional to the surface area A of the slab and the instantaneous absolute temperature change per unit thickness dT/dx of the material. The constant of proportionality C_T is known as the thermal conductivity of the material. The minus sign means that heat flow is in a direction of decreasing temperature.

$$dQ/dt = -C_T A dT/dx$$

7.5   The internal energy U of an ideal gas containing N molecules is proportional to the product of N and the absolute temperature T. The constant of proportionality is 3k/2 where k is Boltzmann's constant.

$$U = (3/2)NkT$$

7.6   In an idealized heated solid called a cavity radiator, the energy radiated from the cavity interior per unit area (called total cavity radiancy, R_C) is proportional to the fourth power of the absolute temperature T. The constant of proportionality σ is called the Stefan-Boltzmann constant.

$$R_C = \sigma T^4$$

8. Temperatures and Conversions

C^0 is the symbol for degrees Celsius, F^0 is the symbol for degrees Fahrenheit and K^0 means degrees Kelvin (Absolute).

8.1 Water freezes at 0 C^0 at standard atmospheric pressure.

8.2 Water boils at 100 C^0 at standard atmospheric pressure.

8.3 The triple point (existing simultaneously as a gas, liquid and solid) of water occurs at a temperature of 273.16 K^0 and atmospheric pressure of 611.73 Pascals (Newtons per square meter).

8.4 $C^0/100 = (F^0 - 32)/180$

8.5 $K^0 = C^0 + 273.16$

Electromechanical Laws

1. Maxwell's Equations:

1.1 The source of the electric field (**E**) is charge density ρ_q. $\nabla = (\partial/\partial x, \partial/\partial y, \partial/\partial z)$ is the normal vector operator, (•) is the normal vector scalar product and ε_0 is a constant called the permittivity of free space. This law is also known as Gauss's law for electricity. The differential form is

$\nabla \bullet \mathbf{E} = \rho_q/\varepsilon_0$

The integral form is

$\varepsilon_0 \oiint \mathbf{E} \bullet \mathbf{n} dS = q$

where $\oiint$ means integration over the closed surface S, **n** is a unit vector normal to S enclosing the charge q.

1.1.1 Maxwell's first equation and may be used to derive Coulomb's law which states that the force between two charges is proportional to the product of the two charges and inversely proportional to the square of the distance between their charge centers. The force is in a direction on a line drawn between the two charges q_1 and q_2 denoted by the unit vector $\mathbf{r}_u$. $K_C = 1/(4\pi\varepsilon_0)$ will be called Coulomb's constant.

$$\mathbf{F} = \mathbf{r}_u K_C q_1 q_2 / r^2$$

giving rise to the electrical potential energy, V between q_1 and q_2

$$V = K_C q_1 q_2 / r$$

If the charges are both positive or both negative, the force is repulsive (like charges repel one another), otherwise the force is attractive (unlike charges attract one another).

1.2 The source of the magnetic field **B** is zero. This is Maxwell's second equation. This also means that magnetic fields always exist in closed loops and magnetic monopoles do not exist. This law is also known as Gauss's law for magnetism. The differential form is

$$\nabla \bullet \mathbf{B} = 0$$

The integral form is

$$\oiint \mathbf{B} \bullet \mathbf{n} dS = 0$$

where $\oiint$ means integration over any closed surface S, **n** is a unit vector perpendicular to the surface, S.

1.3 Ampere's law is also known as Maxwell's third equation. Electrical current density $\mathbf{J}_i$ and/or dynamic electric fields, $\partial \mathbf{E}/\partial t$ give rise to circulating magnetic fields (**B**). μ_0 is known as the permeability constant of free space. The differential form is

$$\nabla \times \mathbf{B} = \mu_0 \mathbf{J}_i + \mu_0 \varepsilon_0 \partial \mathbf{E}/\partial t$$

where $\nabla = (\partial/\partial x, \partial/\partial y, \partial/\partial z)$ is the Del vector operator and **x** is the vector cross product. The integral form is

$$(1/\mu_0) \oint \mathbf{B} \bullet \mathbf{ds} = i$$

where $\oint$ means integration over a closed line s, circulating around the electrical current, i. **ds** is an infinitesimal vector line element of s, that **B** circulates through. **B** is perpendicular to the direction of the electrical current i.

1.4   Faraday's law is also known as Maxwell's fourth equation. Dynamic magnetic fields, $(\partial \mathbf{B}/\partial t)$ give rise to circulating electric fields (**E**). The differential form is

$$\nabla \times \mathbf{E} = -\partial \mathbf{B}/\partial t$$

where $\nabla = (\partial/\partial x, \partial/\partial y, \partial/\partial z)$ is the normal Del vector operator and **x** is the vector cross product. The integral form is

$$\oint \mathbf{E} \bullet d\mathbf{s} = -\iint (\partial \mathbf{B}/\partial t) \bullet \mathbf{n} dS = -\partial \Phi / \partial t$$

where $\Phi = \iint \mathbf{B} \bullet \mathbf{n} dS$ is called the magnetic flux in which **B** penetrates the surface area S. **n** is a unit vector perpendicular to the surface area S.

2. The Lorentz Force:

The force **F** on a charge q moving with velocity **v** by an external electric field **E** and by an external magnetic field **B** and **x** is the normal vector cross product.

$$\mathbf{F} = q\mathbf{E} + q\mathbf{v} \times \mathbf{B}$$

3. Electromagnetic Wave Equations:

When there is no charges or currents, as in the vacuum of matter free space, Maxwell's equations yield a wave equation that is satisfied by both the electric field **E** as well as the magnetic field **B**. These equations yields the precise description of induced electromagnetic fields.

3.1 $\nabla^2 \mathbf{E} - \partial^2 \mathbf{E}/(c^2 \partial t^2) = 0$ and

3.2 $\nabla^2 \mathbf{B} - \partial^2 \mathbf{B}/(c^2 \partial t^2) = 0$

where $\nabla^2 = \nabla \cdot \nabla = \partial^2/\partial x^2 + \partial^2/\partial y^2 + \partial^2/\partial z^2$, t is the time and c is the speed of light in vacuum. Note that if one utilizes the gradient operator, $\Box$ defined as $\Box = (\partial/\partial x, \partial/\partial y, \partial/\partial z, \partial/\partial(ict))$ then, the Dalembertian operator, $\Box^2 = \Box \cdot \Box = \partial^2/\partial x^2 + \partial^2/\partial y^2 + \partial^2/\partial z^2 - \partial^2/(c^2 \partial t^2)$ makes the electromagnetic wave equations 3.1 and 3.2 simplify to

3.1.1 $\Box^2 \mathbf{E} = 0$ and

3.2.1 $\Box^2 \mathbf{B} = 0$

Conservation Laws

1. Conservation of energy: A system's total energy, E_T is the same both before (B) and after (A) any energy transformation.

$$(E_T)_B = (E_T)_A$$

2. Conservation of momentum: A system's total momentum, p_T is the same both before and after any energy transformation.

$$(p_T)_B = (p_T)_A$$

3. Conservation of angular momentum: A system's total angular momentum, L_T is the same both before and after any energy transformation.

$$(L_T)_B = (L_T)_A$$

4. Conservation of charge: A system's total charge, Q_T is the same both before and after any energy transformation.

$(Q_T)_B = (Q_T)_A$

5. Conservation of baryon number: A system's baryon number, N_B is the same both before and after any energy transformation. Baryons are composed of quarks. Quarks have baryon number +1/3. Anti-quarks have baryon number −1/3.

$(N_B)_B = (N_B)_A$

6. Conservation of lepton number: A system's lepton number, N_L is the same both before and after any energy transformation.

$(N_L)_B = (N_L)_A$

7. For any energy system, another related energy system predicted by the simultaneous operations of time reversal, charge conjugation (signs of all

charges involved are reversed) and space reversal (mirror image or parity) is also possible. This is called CPT for short. Below, E_T is the total energy of a system and BCPT means before the CPT operation and ACPT means after the CPT operation.

$(E_T)_{BCPT} = (E_T)_{ACPT}$

Basic Units

position: (measured with a ruler)

meter = m

mass: (measured with a balance scale)

kilogram = kg

time: (measured with a clock)

second = s

charge: (measured with a voltmeter)

coulomb = coul

Equivalent Units

Force: Newton = nt = kg–m/s^2

Pressure: Pascal = nt/m^2

Energy: joule = nt–m

Inductance: henry = joule–m–s^2/coul^2

Capacitance: farad = coul^2/joule

Basic Physical Constants

Name	Symbol	Value
Speed of light	c	3.00×10^8 m/s
Gravitational Constant	G	6.67×10^{-11} nt–m^2/kg^2
Avogadro's number	N_0	6.023×10^{23} /mole
Universal Gas Constant	R	8.32 joules/(mole–K^0)
(Planck's constant)/2π	$\hbar$	1.055×10^{-34} joule–s
Planck length	$L_P = (\hbar G/c^3)^{1/2}$	1.616×10^{-35} m
Planck time	$T_P = (\hbar G/c^5)^{1/2}$	5.391×10^{-44} s
Planck mass	$M_P = (\hbar c/G)^{1/2}$	2.177×10^{-8} kg
Boltzmann's constant	k	1.38×10^{-23} joules/(molecule–K^0)
Stefan-Boltzmann constant	σ	5.67×10^{-8} joules/m^2/(K$^0)^4$
Permeability constant	μ_0	1.26×10^{-6} henry/m
Permittivity constant	ε_0	8.85×10^{-12} farad/m

Name	Symbol	Value
Electron charge	q_e	-1.6022×10^{-19} coul
Electron rest mass	m_e	9.11×10^{-31} kg
Proton rest mass	m_p	1.67239×10^{-27} kg
Neutron rest mass	m_N	1.6747×10^{-27} kg
Coulombs constant	$1/(4\pi\varepsilon_0)$	8.99×10^9 nt-m^2/coul^2

Basic Elementary Particles

Preliminary Particle Descriptors

1. Family Names – Particles belong to functional families having a set number of family members. For example, the gluon family has eight members and they function to provide the strong nuclear force that hold quarks together. Individual particles have both a historical name and a symbol. For example, an electron has the symbol e^-.

2. Color – Quarks can either be red, green or blue (r,g,b). Anti-quarks can either be –red, –green or –blue (–r, –g, –b). This is similar to charge coming in two types, the minus (–) and the plus (+) type.

3. Charge – measured in units of positive electronic charge or the charge on a positron (anti–electron). The charge magnitude of a negative electron (e^-) or a positive positron (e^+) are equal. An anti-particle has the opposite charge as the particle.

4. Spin – Axial angular momentum measured in units of Planck's constant divided by 2π and denoted by $\hbar$. Quantum Spin is specified as positive, but it is understood that quantum mechanically, it can either be positive (parallel) or negative (anti–parallel) to any given direction. Fermions (matter particles) have half integral values of $\hbar$. Bosons (force field particles) have integral values of $\hbar$.

5. Helicity – Helicity is also given in terms of $\hbar$ and may be thought of as the component of the particle's spin in the direction of the particle's velocity vector. The helicity of particles moving at the velocity of light is different than the helicity of particles that do not. Particles moving at the velocity of light, c such as photons, must have zero rest mass and there is no coordinate system for which its velocity is zero. Thus, the component of a photon's spin ($\hbar$) along its velocity vector is the same as its spin orientation, either $+\hbar$ or $-\hbar$ since it cannot be observed at rest. Thus, a photon has an intrinsic helicity the same as

its intrinsic spin. On the other hand, particles with non-zero rest mass have non–intrinsic helicity dependent on the observer since their spin can be observed when they are at rest and their spin components in the direction of motion must have a quantum difference of $+\hbar$. For example, the weakons, responsible for the electroweak forces, with non-zero rest masses and spin of $\hbar$ have helicity of either $-\hbar$, 0, or $+\hbar$. A particle and its anti-particle have opposite helicity.

6. Rest Mass – Measured in either Proton rest masses (Mp) or millions of electron volts (Mev). An electron volt (1.602×10^{-19} joules) is the kinetic energy an electron gains by being propelled a distance of one meter by an electrical field of strength, one volt per meter. The equivalent energy of a proton at rest is 938 Mev. The reason rest mass can be measured in terms of energy is because of Einstein's famous equation $E_0 = m_0 c^2$ which relates rest mass, m_0 to rest mass energy, E_0 by a constant, being the square of the speed of light, c^2.

7. Field Energy – Force fields are caused by corresponding field particles having integral values of $\hbar$ (called bosons). Matter particles having half integral values of $\hbar$ (called fermions) are influenced by force fields caused by their interaction with the corresponding boson. The four force fields are strong nuclear (gluons), electroweak (weakons), electromagnetic (photons) and gravitational (gravitons).

Anti-Particle Properties

All particles have an anti-particle. The anti-particle has the opposite charge of the particle. The anti-particle has the opposite helicity of the particle. The anti-particle of a non-zero rest mass particle having zero charge, and having a spin of one $\hbar$ and zero helicity is the particle itself. The anti-particle has the same mass as the particle. A particle and its anti-particle (that is not itself) annihilate one another upon contact in a burst of other energetic particles.

Matter Energy Particles

All material energy is composed of fundamental matter particles experimentally observed to exist as three energy families (UP, CHARMED, TOP) of four fermions each, in its simplest representation. Two of the fermions are light and are called leptons and two of the fermions are heavy and are called quarks. One of the leptons carries a negative electronic charge, the other has no charge.

Origin of the UP Family

The nuclei of atoms are composed of neutrons and protons. A neutron consists of two (red and blue) down quarks, ($d_R^{-1/3}$, $d_B^{-1/3}$, $u_G^{2/3}$) and one (green) up quark, . A proton consists of two (red and blue) up quarks, and one (green) down quark, ($u_R^{2/3}$, $u_B^{2/3}$, $d_G^{-1/3}$). Any other cyclic permutation of red, green or blue colored quarks in neutrons or protons is possible. The proton is stable. An isolated neutron, n is unstable and will decay into a proton, p electron, e^- and an electron anti-neutrino, $\acute{u}_e$. The net effect is that one of the down quarks of the neutron will change into an electron, anti-neutrino and an up quark. This effectively transformed the internal structure of a neutron ($d_R^{-1/3}$, $d_B^{-1/3}$, $u_G^{2/3}$) into that of a proton ($u_R^{2/3}$, $u_B^{2/3}$, $d_G^{-1/3}$). The up quark has a charge of 2/3 e^+ while the down quark has a charge of $-1/3$ e^+. Thus a proton has a net charge of e^+ while the neutron has a net charge of 0. The UP family making up neutrons and protons consist of four family members which are the up quark, down quark, electron and its anti-neutrino. There are two other four member families. The TOP family has the highest rest mass energy particle members. The CHARMED family has intermediate rest mass energy particle members. The UP family has the lowest rest mass energy particles. Each family maintains the same relationships between its members.

The UP Family

The UP family consists of an up quark, $u^{2/3}$, a down quark, $d^{-1/3}$, electron, e^-, with its electron anti-neutrino, $\acute{\nu}_e$. The quarks can either be red, blue or green. The up quark has a charge of 2/3 e^+ while the down quark has a charge of $-1/3$ e^+. The electron has a rest mass energy of .511 Mev. These particles have the lowest rest mass energy and represent the ground state rest mass energy of the matter families. All UP fermion family members have a spin of $\frac{1}{2}\hbar$ and a helicity of plus or minus $\frac{1}{2}\hbar$.

The CHARMED Family

The CHARMED family consists of a charmed quark, $c^{2/3}$, a strange quark, $s^{-1/3}$, muon, μ^- with its muon anti-neutrino, $\acute{\upsilon}_\mu$. The quarks can either be red, blue or green. The charmed quark has a charge of 2/3 e^+ while the strange quark has a charge of $-1/3$ e^+. The muon has a rest mass energy of 105.66 Mev. These particles have intermediate energy and represent a higher rest mass energy state than the UP family. All CHARMED fermion family members have a spin of $\frac{1}{2}\hbar$ and a helicity of plus or minus $\frac{1}{2}\hbar$.

The TOP Family

The TOP family consists of a top quark, $t^{2/3}$, bottom quark, $b^{-1/3}$, tauon, τ^- with its tauon anti-neutrino, $\acute{\upsilon}_\tau$. The quarks can either be red, blue or green. The top quark has a charge of 2/3 e^+ while the bottom quark has a charge of –1/3 e^+. The tauon has a rest mass of 1784.2 Mev. These particles have the highest rest mass energy state and represent a higher energy state than that of the CHARMED family. All TOP fermion family members have a spin of ½$\hbar$ and a helicity of plus or minus ½$\hbar$.

Field Energy Particles

Gluon Family

Gluons ($g_1 - g_8$) are responsible for the strong force field between the three colored (red, green and blue) quarks making up protons and neutrons, of which all nuclei are composed. There are eight different gluons. Gluons carry color combinations (r, g, b, –r, –g, –b) and compose the gluon field holding quark trios together in protons and neutrons. Gluons have a spin of $\hbar$. Gluons have zero rest mass and therefore move at c, the velocity of light. Thus, gluons have helicity of either plus or minus $\hbar$.

Photon Family

Photons (γ) are responsible for the electromagnetic forces which act between charges. Photons have no color and no charge. Photons have a spin of $\hbar$. Photons have no rest mass and move at the velocity of light. Thus, photons have helicity of either plus or minus one $\hbar$. The positive helicity photon is the anti-photon of the negative helicity photon. While in flight, photons have mass, energy and momentum.

The Weakon Family

Weakons give rise to the electroweak force field responsible for radioactive decay. Recall that a neutron is composed of two down quarks and one up quark. The decay of an isolated neutron is an example of radioactive beta (electron) decay in which one of the down quarks in a neutron decays into a weakon (the omega minus) which then decays into an up quark, electron and anti-neutrino. The net effect is that a neutron decays into a proton, electron and anti-neutrino. There are three different weakons, the omega minus (Ω^-), omega zero or zeta (Z^0) and the omega plus (Ω^+). These weakons have no color and carry charges of e^-, 0, e^+ respectively. Weakons have a spin of $\hbar$ and each can have helicity of $-\hbar$, 0 or $+\hbar$. Weakons have rest masses of 85 Mp, 260 Mp and 85 Mp respectively. Anti-weakons have opposite charges and helicities as the corresponding weakons.

The Meson Families

Mesons give rise to the forces between baryons (quark trios). Mesons are not elemental but are composed of quark anti-quark pairs (combos taken from any of the three families of quarks) and are mentioned here for completeness. Obviously, there are many families of mesons, and the pi meson family (pions) are responsible for forces between nucleons (either neutrons or protons). Pions will be presented next as an example.

The Pi Meson (Pion) Family

The Pions are responsible for forces between nucleons (either neutrons or protons) and are composed of quark anti-quark pairs. The pi minus (π^-) is composed of a down quark with a charge of $-1/3$ e^+ and an anti-up quark with a charge of $-2/3$ e^+ for a total charge of e^-. The pi zero (π^0) is a mixture of an up quark and an anti-up quark, with a down quark and an anti-down quark. The pi plus (π^+) is composed of an up quark with a charge of $+2/3$ e^+ and an anti-down quark with a charge of $+1/3$ e^+ for a total charge of e^+. These pions have no color and carry charges of e^-, 0, e^+ respectively. Pions have a spin of 0 and each has helicity of 0. The charged pions have rest masses of 139.57 Mev, while the pi zero has a rest mass of 134.96 Mev. The anti-pi minus is the pi plus. The anti-pi plus is the pi minus. The anti-pi zero is the pi zero itself.

Graviton Family

Gravitons (G_- and G_+) are responsible for the gravitational force fields which act between masses. Gravitons have no color and no charge.

The G_- graviton has a spin of $2\hbar$ and a zero rest mass. It moves at the velocity of light and thus, its helicity is $-2\hbar$ or $+2\hbar$. It is assumed to have negative mass in flight while being exchanged between any two positive masses or any two negative anti-masses. This is because the gravitational potential energy between two positive masses or two negative masses is negative.

Because of a new scientific theory called "Nature of the First Cause", in which positive matter is gravitationally repelled by negative anti-matter, the G_+ graviton is postulated to exist. It also has a spin of $2\hbar$. It is assumed to have a zero rest mass and moves with the velocity of light and has positive mass in its flight between negative anti-matter and positive matter. Thus, the G_+ graviton also has helicity of $-2\hbar$, or $2\hbar$. The G_+ gravitons fill up all spacetime and are responsible for the force of repulsion between negative anti-matter and positive matter. By the "First Cause" theory, it makes up the repulsive gravitational field which is responsible for the accelerated expansion of distant positive matter in the universe (galaxies not in the local group).

The Higgs Family

There are two Higgs bosons (H_L and H_H) called the light Higgs boson, H_L of the unified electroweak theory and the heavy Higgs boson, H_H of the grand unified theory. The heavy Higgs boson, makes up the Higgs field and permeates all spacetime. This field is responsible for assigning masses to all the fundamental particles. The light Higgs boson is responsible for assigning the masses to the weakons. Both Higgs bosons have a spin of zero (0), and thus they both have a helicity of zero. Both Higgs bosons have non–zero rest masses with the light Higgs rest mass at roughly 10^5 Mev and the heavy Higgs rest mass of about 10^{17} Mev.

Complete Set of Particles

All matter particles which have been discovered are combinations of the above elementary matter particles. All the known force fields consists of varying energy and intensity of the above force field particles.

The Hadrons (consisting of quarks) which are matter particles that have been discovered now number over two hundred, exceeding the number of known elements.

References

Al-Khalili, Jim, *Quantum, A Guide for the Perplexed*, United Kingdom, Weidenfeld & Nicolson, 2003

Ames, Joseph Sweetman & Murnaghan, Francis D., *Theoretical Mechanics An Introduction to Mathematical Physics*, New York, Dover Publications, Inc., 1957

Atkins, K. R., *Physics*, New York, John Wiley & Sons, Inc., 1965

Bennett, Jeffrey & Donahue, Megan & Schneider, Nicholas & Voit, Mark, *The Cosmic Perspective*, New York, Addison Wesley, 2004

Bergmann, Peter Gabriel, *Introduction to the Theory Of Relativity*, New York, Dover Publications, Inc., 1976

Blass, Gerhard A., *Theoretical Physics*, New York, Appleton-Century-Crofts, 1962

Bova, Ben, *The Fourth State of Matter*, New York, New American Library, Inc., 1974

Born, Max, *Einstein's Theory of Relativity*, New York, Dover Publications, Inc., 1962

Breithaupt, Jim, *Cosmology*, Blacklick, OH, McGraw-Hill, 1999

Chrien, Robert E., *Focus on Physics Nuclear Physics*, New York, Barnes & Noble, 1972

Davies, Paul, *The New Physics*, New York, Cambridge University Press, 1996

De Broglie, Louis, *matter and light*, New York, Dover Publications, Inc., 1939

Einstein, Albert, *Builders of the Universe*, Los Angeles, CA, U. S. Library Association, Inc., 1932

Einstein, Albert, & Lorentz, H. & A., Minkowski, H., & Weyl, H., *The Principle of Relativity*, New York, Dover Publications, Inc., 1952

Einstein, Albert, *Relativity The Special and General Theory*, New York, Crown Publishers, Inc., 1961

Fermi, Enrico, *thermodynamics*, New York, Dover Publications Inc., 1956

Feynman, Richard P., *QED The Strange Theory of Light and Matter*, Princeton, New Jersey, Princeton University Press, 1988

Feynman, Richard P., *Six Not So Easy Pieces*, New York, Basic Books, 1997

Frankel, Theodore, *Gravitational Curvature An Introduction to Einstein's Theory*, San Francisco, W. H. Freeman and Company, 1979

Gamow, George, *Gravity*, New York, Dover Publications, Inc., 2002

Goldstein, Herbert, *Classical Mechanics*, London, Addison-Wesley Publishing Company, Inc., 1950

Greene, Brian, *The Elegant Universe: Superstrings, Hidden Dimensions, and the Quest for the Ultimate Theory*, New York, W. W. Norton, 1999

Guth, Alan H., *The Inflationary Universe: The Quest for a New Theory of Cosmic Origin*, Perseus Books, 1997

Halliday, David & Resnick, Robert, *Physics For Students of Science and Engineering*, New York, John Wiley & Sons, Inc., 1962

Hawking, Stephen & Penrose, Roger, *The Nature of Space and Time*, New Jersey, Princeton University Press, 1996

Heisenberg, Werner Karl, *The Nature of Elementary Particles*, in Physics Today, Page 39, March 1976

Kaku, Michio, *Hyperspace*, New York, Anchor Books, 1995

Kaku, Michio, *Parallel Worlds*, New York, Anchor Books, 2006

Kaplan, Irving, *Nuclear Physics*, Reading, Massachusetts, Addison-Wesley Publishing Company, Inc., 1962

McMahon, David, *quantum mechanics demystified*, New York, McGraw Hill, 2005

Messiah, Albert, *Quantum Mechanics*, New York, Dover Publications, Inc., 1999

Musser, George, *Growing Pains*, Scientific American, Page 32, July 2004

Ostriker, Jeremiah P., & Steinhardt, Paul J., *The Quintessential Universe*, Scientific American, Page 46, January 2001

Park, David, *Introduction to the Quantum Theory*, New York, McGraw-Hill Book Company, 1964

Peebles, P. J. E., *Principles of Physical Cosmology*, Princeton, New Jersey, Princeton University Press, 1993

Penrose, Roger, *The Road To Reality, A Complete Guide to the Laws of the Universe*, New York, Vintage Books, 2004

Powell, John L. & Crasemann, Bernd, *Quantum Mechanics*, Reading, Massachusetts, Addison-Wesley Publishing Company, Inc., 1961

Ridpath, Ian, *The Illustrated Encyclopedia of the Universe*, New York, Watson-Guptil Publications, 2001

Riggs, Shelton, *An Alternative Lorentz Invariant, Relativistic Wave Equation,* Version 7.7, El Paso, Texas, www.co.el-paso.tx.us/clerk/deedsearch.htm, instrument number 20060091478, El Paso County Courthouse, 2006

Sears, Francis W., & Zemansky, Mark W. & Young, Hugh D., *College Physics*, Menlo Park, California, Addison-Wesley Publishing Company, 1986

Segre, Emilio, *Nuclei and Particles*, New York, W. A. Benjamin, Inc., 1964

Shortley, George & Williams, Dudley, *Elements of Physics For Students of Science and Engineering*, Englewood Cliffs, New Jersey, Prentice-Hall, Inc., 1965

Van Heuvelen, Alan, *Physics, A General Introduction*, Boston, Little, Brown and Company, 1982

Weinberg, Steve, *Dreams of a Final Theory: The Search for the Fundamental Laws of Nature*, New York, Pantheon Books, 1992

Weld, LeRoy D., *A Textbook of Heat*, New York, The Macmillan Company, 1948

Weyl, Hermann, *Symmetry*, Princeton University Press, 1952

Young, Hugh D., *Statistical Treatment of Experimental Data*, McGraw-Hill Co., Inc., 1962

INDEX

A

absolute temperature, 130, 131
accelerated expansion, 157
acceleration, 22, 115, 117, 118, 121
Acceleration, 115
action, 121
angular momentum, 61, 118, 123
Angular momentum, 118
anti-gravitons, 89
anti-matter, 157
anti-neutrino, 150
anti–neutrino, 112, 151, 152, 154
anti-parallel, 67
anti-particle, 145, 147, 148
anti–particle, 113
Anti-Particle, 148
Anti-weakons, 154
atmospheric pressure, 132
attractive, 75

Average mass density, 118
axial angular momentum, 5
Axial angular momentum, 146
axial spin, 68

B

beryllium, 101
Boltzmann's constant, 131, 143
Bose-Einstein Statistics, 105
boson, 112
Boson, 105, 109, 110
Bosons, 105, 146
bottom quark, 112, 152

C

Celsius, 132
charge, 113
Charge, 79
charge density, 119, 133
Charge density, 119
charges, 112
CHARMED family, 150, 151, 152
charmed quark, 85, 151

Cold Fusion, 96

cold fusion reactors, 104

Color, 145

color charge, 1

Color Confinement, 18

color field, 11

color force, 77

color forces, 30

Color Forces, 15

color spring constant, 42

colorless, 6

Conservation of angular momentum, 139

Conservation of baryon number, 140

Conservation of charge, 139

Conservation of energy, 139

Conservation of lepton number, 140

Conservation of momentum, 139

coordinate system, 114, 146

copper, 100

coulombs constant, 40

Coulomb's law, 134

Coulomb's Law, 105

current, 120, 136

D

Dalembertian operator, 138

De Broglie, 126

De Broglie's, 126

del operator, 48

Del Operator, 106

deuterium, 68, 103

direction, 23

down quark, 5, 6, 111, 112, 149, 150, 155

E

E-Cat, 99

eigenvalue, 50, 57, 106, 108

eigenvalues, 52

Einstein, 25, 161

Einsteinian, 114, 115, 116, 117, 118, 123, 128

Einstein's, 47

Electric Charge, 7

electric field, 120, 133, 137, 138

electric fields, 136

electrical coulomb forces, 30

electrical current, 119, 136

Electrical current, 119
electrical current density, 120
Electrical current density, 119, 135
electromagnetic, 109, 125, 148, 153
Electromagnetic, 137
electron, 108, 109, 111, 145, 149, 150, 154
Electron, 144
electron anti-neutrino, 149, 150
electron anti–neutrino, 111
electron neutrino, 82, 83, 84
electron volt, 106, 147
electroweak, 147, 148, 153, 157
Elementary Particles, 145
energy, 112, 117, 122, 123, 128, 129, 131, 134
   field, 112
   internal, 129, 131
   kinetic, 128
   potential, 122, 134
   total, 117, 123, 139
   transformation, 139
Energy eigenvalues, 123
entropy, 129

F

Family Names, 145

Farenheit, 132

Fermi-Dirac Statistics, 106

fermion, 111

Fermion, 107, 108, 109

fermions, 62, 148, 149

Fermions, 106, 107, 109, 146

Feynman Diagram, 107

Field Energy, 148, 152

field particles, 148, 158

fields, 135, 136

   electric, 120, 133, 135, 137

   gravitational, 122

   magnetic, 134

force, 21, 22, 23, 48, 109

force field particles, 146

force fields, 148, 156, 158

Force Particles, 88

G

galaxies, 157

Gauss, 133, 135
Gauss's law, 11
gluon, 112
Gluon Family, 90
gluon family., 88
gluons, 37, 112, 148, 152
Gluons, 152
gravitational, 53, 122, 123, 148, 156, 157
Gravitational, 122, 143
graviton, 95, 112, 156
graviton family, 89
Graviton Family, 156
gravitons, 148, 156
Gravitons, 156

H

half integral, 107
Hamiltonian operator, 49, 50, 56
Hamiltonian Operator, 49, 108
hardware, v
heavy Higgs, 157
Heisenberg, 124
Heisenberg uncertainty principle, 125

helicity, 148
Helicity, 146
HELICITY, 108
helium, 101
Higgs bosons, 157
Higgs Family, 157
Hooke's law, 37
Hooke's law, 19

I

Instantaneous mass density, 118
integral, 23
intrinsic helicity, 146

K

Kelvin, 132
kinetic energy, 23, 24, 50, 56, 106, 128, 147

L

Laplacian, 48, 52, 108
Laplacians, 51
law of heat conduction, 130
Laws, 165

Laws of Special Relativity, 126
Laws of Thermodynamics, 129
length, 127, 143
LENR, 97
leptons, 111, 149
light Higgs, 157
local group, 157
Lorentz, 46, 47
Lorentz Force, 137

M

magnetic field, 134, 137, 138
magnetic fields, 135
mass, 113, 116, 117, 120, 121, 122, 126, 128, 142, 143, 144
Mass, 122, 123
Mass density, 118
material energy, 149
matter, 160
MATTER
   particles, 111, 113
matter particles, 146, 149, 158
Matter particles, 79

Matter Particles, 81

Maxwell's, 133, 134, 135, 136, 137

Maxwell's equation, 12

Maxwell's equations, 137

mc², 23, 24, 25, 26, 27, 39, 40

measurement, 124, 125

mechanical energy, 128

Meson Families, 154

Mesons, 154

Mev, 147, 151, 152, 155, 157

MEV, 109

muon, 85, 112, 151

muon neutrino, 82, 85, 86

N

negative anti-matter, 156

negative mass, 156

negative masses, 156

neutron, 1, 5, 149, 153

Neutron, 144

neutrons, 112

Neutrons, 4

newton, 142

Newton, 109

Newtonian, 21, 114, 115, 116, 117, 122, 128, 129

Newton's, 121, 122

Nickel, 98

O

omega minus, 89, 92, 112, 154

omega plus, 89, 92, 112, 154

orbital angular momentum, 53, 60, 64

oscillator potential, 27

P

Palladium, 98, 103

Particle Generator, 80

particles, 49, 50, 56, 57

Pascal, 142

Pascals, 132

Pauli Exclusion Principle, 109

perfect gas law, 130

photon, 69, 112, 125

Photon Family, 91

photons, 50, 53, 57, 107, 146, 148, 153

Photons, 153

pions, 112

Pions, 6, 154, 155

Planck Acceleration, 110

Planck Boson Spin, 110

Planck Energy, 110

Planck Fermion Spin, 110

Planck Force, 110

Planck Frequency, 110

Planck Length, 111

Planck Mass, 111

Planck Momentum, 111

Planck Time, 111

Planck's constant, 5, 48, 105, 107, 124, 125, 126, 143, 146

Planck's Constant, 110

position, 114, 118, 124, 125

Position, 114, 142

positive matter, 156

positron, 7, 145

potential, 48, 49, 56

potential energy, 24, 49, 56, 156

pressure, 118, 129, 130, 132

Pressure, 118

Proof, 79

proton, 1, 5, 147, 149, 154

Proton, 144

protons, 112

Protons, 4

prototype reactor, 99

Q

quantum mechanics, v

quantum number, 53

quantum numbers, 109

quark, 111, 112

quark anti-quark pairs, 154

quark color, 4

Quark Families, 82

quarks, 1, 111, 112, 140, 145, 149, 150, 151, 152, 153, 154, 155, 158

R

radioactivity, 108, 109

rationalized Planck's constant, 61

relativistic, 21, 22, 23, 47, 49, 53

relativistic harmonic oscillator, 21, 25

Relativistic Wave Equation, 46, 47
relativistic wave mechanics, 53
repulsive, 74
rest mass, 24, 79, 127, 128, 146, 147, 148, 150, 151, 152, 153, 155, 156, 157
Rest Mass, 147
rest mass energy, 24, 147, 150, 151, 152

S

Scientific American, 163
silver, 103
simple harmonic oscillator, 19
special relativity, 25
Speed of Light, 111
spin, 123
Spin, 146
spin angular momentum, 105, 107, 108
spin-orbit coupling, 20
Spin-Orbit Coupling Forces, 64
spin-spin coupling, 20
Spin-Spin Coupling Forces, 62
square well potential, 27
STANDARD, 111

standard model, 37
stationary energy states, 50, 57
Stefan-Boltzman, 143
Stefan-Boltzmann constant, 131
strange family, 82
Strange Family, 85
strange quark, 85, 112, 151
strong nuclear, 145, 148

T

tauon, 112, 152
tauon neutrino, 82, 87, 88
theory of relativity, 47
time, 114, 115, 117, 119, 127, 138, 140, 142, 143
time derivative, 22
top family, 82
Top Family, 87
TOP family, 150, 152
top quark, 152
total energy, 24, 48, 50, 57, 108, 123, 139, 140
total energy of the proton, 33
transmutation, 99
triple point, 132

tritium, 71
tritium nuclei, 73

U

Uncertainty Principle, 124
Unlimited Number of Primes, 79
up family, 82
Up Family, 83
UP family, 150, 151
UP Family, 149, 150
up quark, 6, 112, 149, 150, 153, 155
up quarks, 5

V

vector, 48, 49, 56, 106, 115, 116, 118, 119, 122, 133, 134, 135, 136, 137
velocity, 115, 116, 117, 118, 121, 126, 127, 128, 137
Velocity, 114
velocity vector, 108, 146

W

Water boils, 132

Water freezes, 132
wave equation, 137
Wave Equation, 123
   electromagnetic, 137
   quantum mechanical, 123
wave function, 48, 50, 57, 106, 108
Wave Function, 113
weak force, 89
weakon, 154
Weakon Family, 92
weakons, 112, 147, 148, 154, 157
Weakons, 153

Z

zero rest mass, 47, 107, 147
zeta, 112, 154
zeta zero, 93

www.ingramcontent.com/pod-product-compliance
Lightning Source LLC
Chambersburg PA
CBHW071424170526
45165CB00001B/393